C0-CDD-658

When Ideology Trumps Science

When Ideology Trumps Science

Why We Question the Experts on Everything from Climate Change to Vaccinations

Erika Allen Wolters and Brent S. Steel

An Imprint of ABC-CLIO, LLC
Santa Barbara, California • Denver, Colorado

Library of Congress Cataloging in Publication Control Number: 2017034264

ISBN: 978–1–4408–4983–1 (print)
978–1–4408–4984–8 (ebook)

22 21 20 19 18 1 2 3 4 5

This book is also available as an eBook.

Praeger
An Imprint of ABC-CLIO, LLC

ABC-CLIO, LLC
130 Cremona Drive, P.O. Box 1911
Santa Barbara, California 93116-1911
www.abc-clio.com

This book is printed on acid-free paper ∞

Manufactured in the United States of America

Contents

List of Tables vii

Preface xi

1 Introduction **1**
Science versus Policy 2
Motivated Reasoning 4
Political and Value Orientations 7
Methods and Approach 12
Attitudes toward Science and Scientists 13
Ideology and Value Orientations 16
Discussion and Book Plan 23
References 24

2 Genetically Modified Organisms (GMOs) **29**
Introduction 29
What Is a GMO? 30
Ideology, Values, and GMO Beliefs 35
Analyses 38
Discussion 46
References 49

3 Childhood Vaccinations **53**
Introduction 53
The Controversy 56
Vaccine Skeptics 60
Ideology, Positivism, and Postmaterialism 62
Analyses 66

Conclusion 77
References 80

4 Climate Change 85
Introduction 85
Climate Change Science and Impacts 86
Direct Impacts of Climate Change 87
Climate Change and the U.S. Public 90
Ideology, Positivism, and Postmaterialism 93
Analyses 98
Conclusion 103
References 105

5 Teen Sex 109
Introduction 109
Human Papillomavirus (HPV) 111
Plan B—Emergency Contraception 114
Abstinence-Only Education 116
Ideology, Positivism, and Postmaterialism 119
Analyses 123
Conclusion 130
References 132

6 Stem Cell Research 137
Introduction 137
History and Politics of ESCs 138
Stem Cells, Ideology, and Values 141
Analyses 143
Conclusion 148
References 150

7 Conclusion: What Is to Be Done? 153
Introduction 153
The Democracy-versus-Technocracy Quandary 154
Ideology, Values, Science, and Policy Issues 156
Science, the Public, and the Policy Process 162
Summary 171
References 173

Index 177

Tables

1.1	Survey Response Bias	13
1.2	Public Self-Assessed Awareness and Issue Salience of Scientific Issues	14
1.3	Public Preferences for the Role of Science and Scientists in the Policy Process	16
1.4	Preferred Role for Science and Scientists in the Policy Process	17
1.5	Value Orientation Indicators	17
1.6	Ideology and the Role of Science and Scientists in Policy	19
1.7	Ideology and Preferred Role of Science and Scientists in Policy	20
1.8	Postmaterialist Values and the Role of Science and Scientists in Policy	20
1.9	Postmaterialist Values and Preferred Role of Science and Scientists in Policy	21
1.10	Positivism Beliefs and the Role of Science and Scientists in Policy	22
1.11	Positivism Beliefs and Preferred Role of Science and Scientists in Policy	23
2.1	West Coast Public and GMO Orientations	39
2.2	Correlation Coefficients for GMO Orientations (Kendall's Tau b)	39
2.3	Political Ideology and GMO Orientations	40
2.4	Postmaterialist Values and GMO Orientations	42

2.5 Belief in Positivism and GMO Orientations 43
2.6 Regression Estimates for GMO Orientations 45
3.1 Comparison of Prevaccine Estimated Average Annual Cases vs. Most Recent Reported Vaccine Cases from 2006, United States 54
3.2 Public Attitudes toward Vaccinations 67
3.3 Correlation Coefficients for Vaccination Orientations 68
3.4 Political Ideology and Vaccination Orientations 70
3.5 Postmaterialist Values and Vaccination Orientations 71
3.6 Positivism Beliefs and Vaccination Orientations 72
3.7 Regression Estimates for Antivaccination Orientations 75
4.1 Public Orientations Concerning Climate Change 99
4.2 Political Ideology and Climate Change Orientations 100
4.3 Postmaterialist Values and Climate Change Orientations 101
4.4 Positivism Beliefs and Climate Change Orientations 101
4.5 Logistic Regression Estimates for Climate Change Orientations 103
5.1 Public Attitudes toward HPV, Plan B, and Abstinence Orientations 124
5.2 Correlation Coefficient (Tau b) for HPV, Plan B, and Abstinence Orientations 125
5.3 Political Ideology and HPV, Plan B, and Abstinence Orientations 126
5.4 Postmaterialist Values and HPV, Plan B, and Abstinence Orientations 127
5.5 Positivism Beliefs and HPV, Plan B, and Abstinence Orientations 129
5.6 Regression Estimates for HPV, Plan B, and Abstinence Orientations 130
6.1 Public Attitudes toward Stem Cell Research 144
6.2 Political Ideology and Stem Cell Orientations 145
6.3 Postmaterialist Values and Stem Cell Orientations 145
6.4 Positivism Beliefs and Stem Cell Orientations 146
6.5 Regression Estimates for Stem Cell Orientations 147
7.1 Political Ideology and Scientific Controversies: Liberals vs. Conservatives 157

7.2 Postmaterialist Values and Scientific Controversies: Postmaterialists versus Materialists 159
7.3 Positivism Beliefs and Scientific Controversies: Low vs. High Levels of Positivism Beliefs 161
7.4 Scientific Autonomy and Integrity 162
7.5 Conditions Affecting the Impact of Scientific Information on Public Policy 163
7.6 Political Ideology and Belief in Positivism 170
7.7 Ideology and Belief in Positivism Combined 170

Preface

The United States is currently at a crossroads. We are now in a political era where facts are fluid and the truth is subjective. This is dangerous territory. Science and empirical facts are the proverbial glue that should transcend ideology and worldviews. When we disregard science in order to construct a reality that fits more into the way we want the world to be rather than the way it is, we risk outcomes that do not adhere to the laws of science. This book provides a sober analysis of how embedded beliefs rather than a lack of scientific knowledge and understanding are creating a cognitive bias toward information that coincides with personal beliefs rather than scientific consensus—and that this antiscience bias exists among liberals as well as conservatives.

Whether we are cognizant or not of the enormous influence science and technology has had on the United States, it has shaped our economy, society, and culture in innumerable ways. Instead of being critical analysts of scientific information, we default instead to cultural constructs of values and worldviews to inform our policies on critical social and environmental issues. The United States has maintained a global leadership role because of our advancements in science and technology. However, at a time when science is even more critical in helping policy makers and the public understand crucial issues like climate change and food production, science is being sidelined for ideological or personal values.

The consequences of ideology trumping science can be devastating. For example, while vaccines exist for many diseases, some parents chose not to vaccinate their children as a result of personal fears and a distrust of the scientific community. A 2010 outbreak of whooping cough in California infected more than 8,000 people, resulting in the hospitalization of over 800 people and the death of 10 infants. In 2015, an outbreak of the measles in Disneyland infected more than 125 people. Both the whooping cough and the measles are vaccine-preventable diseases that have been largely nonexistent in the United States for decades. As these cases demonstrate, individuals who prioritize ideology or personal beliefs above scientific consensus can impinge on society at large—rejecting science has unfortunate results for public health and the environment. The effects of climate change may lead to even more drastic long-term and global consequences for human health, lifestyles, food supplies, and other deleterious impacts.

It is our hope that this book may play a small role in getting people to think about the proper role of science and scientists in society and the policy process, and to reflect on their own values and ideology concerning their acceptance or rejection of scientific information—especially when there is a consensus in the scientific community.

CHAPTER 1

Introduction

The good thing about science is that it's true whether or not you believe in it.

—Neil deGrasse Tyson

In recent decades, it has become clear that science and scientists no longer hold sway as unquestioned authoritative sources of credible information in many high-stakes policy debates. In case after case, whether it is climate change, GMO (genetically modified organism) food, immunization, stem cell research, or birth control, science and the scientists responsible for conducting and communicating the applicable research to decision makers encounter more frequent and direct challenges than ever before. For example, in California the governor recently signed a law, amid much protest from the right and left, eliminating personal exemptions against vaccination of public school children; in Texas, support for abstinence-only sex education remains high despite recent upsurges in sexually transmitted diseases (STDs); and across the nation many Republican candidates for office deny climate change in order to appeal to conservative voters amid evidence that 2015 was the hottest year on record.

Taken together, these challenges highlight the contested nature of contemporary perspectives on science and its proper role in the policy process; the possibility that policy makers will not take science into account when making decisions; and the fact that citizens in developed countries, such as the United States, are increasingly unlikely to hold views consistent with scientific consensus. The objectivity and truth claims of traditional science have come under attack by some on both the left and the right, who see modern science as simply another expression of power that favors certain elites, while also seeking to discredit and marginalize other truth claims, most notably those of religion. Others in this vein classify science, and the practice of scientists, as but another social institution with its own particular social and cultural "processes" that are inevitably replete with politics and values.

This book is an exploration of how values and worldviews overshadow scientific consensus on climate change, GMOs, vaccinations, abstinence-only education, issues related to teen sexual activities, and stem cell research. In each case, those who question the science marginalize scientific agreement in favor of policies that best reflect their personal beliefs and preferences, often at the peril of the environment and public health.

SCIENCE VERSUS POLICY

On March 28, 1979, the Three Mile Island nuclear power plant in Pennsylvania experienced a partial meltdown. Four days later on April 1st, the situation was contained and the crisis was determined to be over. In the months following the meltdown, several government agencies conducted studies on the potential health impacts to residents that found that among the 2 million people potentially exposed, there were no adverse health effects that could be attributed to the exposure to radiology (U.S. Nuclear Regulatory Commission, 2014). In fact, people were exposed to a fraction of the amount of radiation in an X-ray, and "in spite of serious damage to the reactor, the actual release had negligible effects on the physical health of individuals or the environment" (U.S. Nuclear Regulatory Commission, 2014).

The resulting consequences of the meltdown impacted the public perception of nuclear energy safety, particularly regarding public health. Further, Three Mile Island mobilized the antinuclear movement and became the symbol of nuclear instability and danger. In 1986, the Chernobyl disaster in the former Soviet Union confirmed fears over nuclear safety with more than half a million people affected by high levels of radiation and the death of 31 people. Together, these events solidified resistance to nuclear energy, particularly among liberal Democrats, even though many scientific experts often praise the overall safety record of the U.S. nuclear industry (Porter, 2016).

Conservatives are currently leading the opposition to climate change legislation that would curb greenhouse gas (GHG) emissions for the United States, often denying that climate change is even real. Yet 60 percent of Republicans favor building nuclear power plants (Pew Research Center, 2015), a decidedly clean energy source that could offset oil and coal and reduce GHG emissions. Alternatively, only 35 percent of Democrats support building nuclear power plants (Pew Research Center, 2015), although they are the most adamant about climate change policies to reduce GHG emissions. Somewhat ironically, Republicans are more aligned with scientists in their support of nuclear energy, with 65 percent of scientists in the American Association for the Advancement of Science concurring that more nuclear energy facilities should be built (Porter, 2016).

The controversy over nuclear energy illustrates an ongoing disparity between beliefs shared by the majority of scientists and the biases of the public. However, the lack of public support for science policy issues is often less about science and more about worldviews. If there is consensus that science holds truth through rigorous testing of hypotheses and resulting evidence to support or disprove a theory, then science cannot be used like a coat on a cold day—something to take on and off at an individual's discretion. And yet even people who strongly support scientific evidence in one policy issue may strenuously object to the same scientific principles being applied to other policy issues.

In 2015, the Pew Research Center released a study on the similarities and differences between "Public and Scientists' Views on Science and Society." Results from this study found that although both the public and scientists hold science in high regard, there are rather sizable discrepancies in their views regarding several key policy areas (Pew Research Center, 2015). Among some of the findings were an 18 percent gap between the public and scientists on the issue of requiring the measles, mumps, and rubella (MMR) vaccine (68% public compared to 86% scientists). Similarly, on the issue of genetically modified foods, a 51 percent gap exists between the public and scientists: 88 percent of scientists feel they are safe to eat, while only 37 percent of U.S. adults believe they are safe to eat (Pew Research Center, 2015). Further, U.S. adults have difficulty demonstrating what is generally considered scientific consensus. When asked whether the universe was created by "the Big Bang," 52 percent of U.S. adults said that "scientists are divided," while only 42 percent said "scientists generally believe" (Pew Research Center, 2015). The lack of public knowledge of science concerns scientists, with 84 percent saying that the "public doesn't know much about science," in part blaming the media for "oversimplifying the problem" (Pew Research Center, 2015).

The difficulty with this assessment is that we are now living in a time where "information" is abundant. A simple click on a computer can open a plethora of information pertaining to almost any current science-policy problem. A study conducted by the National Science Foundation (NSF) found that in 2014 more than half of Americans used the Internet for scientific and technological information (NSF, 2016). However, while seeking information, people gravitate to information that already adheres to their preformed beliefs. In other words, people selectively look for information that confirms their preconceptions and beliefs surrounding a policy issue. So, it may be that it is not simply a paucity of knowledge that leads to misinformation or scientific illiteracy, but perhaps instead a confirmation bias toward alternative "evidence" that best supports preheld worldviews.

MOTIVATED REASONING

Get your facts first, and then you can distort them as much as you please.

—Mark Twain

In the face of complex policy decisions, logic would dictate that when new scientific information is obtained that contradicts individuals' understanding of an issue, individuals would modify their opinion to reflect their acquired knowledge. This "normative" model regarding decision making implies "a two-step updating process, beginning with the collection of belief-relevant evidence, followed by the integration of new information with the prior to produce and updated judgment" (Taber and Lodge, 2006: 755). However, humans in all of their complexities do not comply with the rules of logic all of the time. So although people "are always constrained in some degree to be accurate, they are typically unable to control their preconceptions, even when encouraged to be objective" (Taber and Lodge, 2006: 756).

Motivated reasoning suggests that emotion-based decision making obstructs the ability of individuals to assimilate new information that is contrary to their preferred policy positions, what is known as disconfirmation bias. People seek information that conforms to their preconceived preferences, enabling a confirmation bias, or an attitude congruence bias, to occur when searching for information. Therefore information obtained is demarcated deliberately to support existing policy preferences, and other nonconforming information is seen as not as accurate or scientifically sound (Taber, Cann, and Kucsova, 2009).

When an individual seeks information about a policy issue, the extent to which he or she acquires and processes information is dependent on being motivated by either an accuracy goal or a directional goal. An accuracy goal is when people seek information that is correct, regardless of individual beliefs about the issue (Nir, 2011). Accuracy goals are driven by what is known as "cold cognition," meaning that information is systematically obtained in a logical, impersonal nature. In contrast, directional goals are "defined as needs to uphold and maintain a desirable conclusion and reject disconfirming information" (Nir, 2011: 506). Directional goals are driven by "hot cognition," that is, an emotion-based process of obtaining information that reiterates preheld beliefs or policy positions.

While several studies have confirmed the role of motivated reasoning in decision making (Redlawsk, 2002; Taber and Lodge, 2006), a unique study conducted by Westen et al. utilized neuroimagery to "study the neural responses of 30 committed partisans during the U.S. Presidential election of 2004" (2006: 1947). Interestingly, the study confirmed prior research regarding motivated reasoning; that is, when subjects were presented with information about their candidate "that would logically lead them to an emotionally aversive conclusion, partisans arrived at an alternative conclusion" (2006: 1955). This "motivated reasoning was not associated with neural activity in regions previously linked to cold reasoning tasks and conscious (explicit) emotion regulation" (2006: 1947). This study therefore confirmed that emotional decision making (motivated reasoning) is related to "hot cognition" and not logical "cold cognition," which drives an individual to seek accurate information, regardless of beliefs.

Just as people select information sources that already conform to their beliefs (e.g., conservatives watch FOX NEWS, while liberals watch MSNBC), it is this same selective bias that hinders the acceptance of new information that conflicts with established preferences and reinforces shared beliefs among groups. This notion of cultural cognition "refers to the tendency of individuals to fit their perceptions of risk and related factual beliefs to their shared moral evaluations of putatively dangerous activities" (Kahan, Jenkins-Smith, and Braman, 2011: 148). In public policy, groups often coalesce around ideological and partisan beliefs. For liberals, views tend to gravitate toward egalitarian and communitarian values, while for conservatives the values are more hierarchical and individualistic (Jost, Nosek, and Gosling, 2008). Put another way, Janoff-Bulman (2009) argues that liberals seek to provide, while conservatives aim to protect. Liberals are focused more on matters of social justice and positive social gains, hold a more open view on alternative lifestyles, and seek inclusivity in society (Janoff-Bulman, 2009). Alternatively, conservatives are more focused on order and structure that comes with restrictions to avoid losses to society (Janoff-Bulman, 2009) and as such are more likely to be religious (Jost, Nosek, and Gosling,

2008). It is this ideological divide that "helps to explain why people do what they do; it organizes their values and beliefs and leads to political behavior" (Jost, 2006: 653).

It is easy to suggest that people simply do not have enough information or do not trust science to explain why public opinions often fall short of understanding scientific consensus. But reality is more complicated than that. In terms of the U.S. public's views regarding science, 79 percent hold a positive feeling about science in society (Pew Research Center, 2015). This was similarly found in an NSF 2016 study where 7 in 10 Americans agreed that scientific benefits are greater than harms, and 9 in 10 agree that more opportunities will be created by science and technology for future generations (NSF, 2016). While the public holds generally positive views about science broadly, issue-specific and scientific knowledge among the U.S. public is less encouraging. However, as Kahan, Braman, and Jenkins-Smith argue:

> The same tendency individuals have to attend to information in a biased way that reinforces their priors could lead them to form biased assessments of the authority and knowledge of putative experts in a manner that fits their predispositions. This process, too, would lead individuals of opposing outlooks to arrive at radically different results when they conjure examples of "expert opinion" on particular issues . . . And, as such scientific consensus cannot be expected to counteract the polarizing effects of cultural cognition because apprehension of it will necessarily occur through the same social psychological mechanism that shape individuals' perceptions of every other manner of fact. (2011: 151–52)

Therefore it is not that Americans do not trust scientists, it is that they are discerning about which scientists and scientific knowledge to view as both credible and accurate.

Americans primarily get their information from TV, newspapers, radio, or the Internet. Of the information distributed, nearly 80 percent of scientists feel communication by the media is often problematic in that it does not do a good job of vetting "well-founded" scientific findings versus junk science (Pew Research Center, 2015). However, with an increasing number of Americans (62%) getting their news from social media sites (like Facebook and Twitter) (Pew Research Center, 2016), information about science and technology through these sites is effectively what amounts to sound bites, requiring people to obtain more in-depth information from other sources, often sources that comply with their preformed opinions. Because people are limited in time and attention, it is easier to seek information that already conforms to beliefs, and become more politically polarized even in light of corrective information about a candidate or policy (i.e., a "backfire effect" or a "boomerang effect") (Hart and Nisbet, 2011; Nyhan and Reifler, 2010).

It is evident that people's beliefs strongly impact their opinions related to science-policy issues. While Americans generally support science in the abstract, they are less committed to scientific findings, particularly related to areas where scientific consensus exists on controversial issues. In these cases, people rely on science that conforms to what they already believe and are more likely to reject information that contradicts these beliefs. Since much of people's belief formation is contingent on their political and value orientations, these variables likely influence policy preferences by potentially contributing to confirmation bias, and help shape an individual's worldview. Indeed, even in the face of overwhelming scientific evidence, it is these embedded factors that filter information that forms policy preferences.

POLITICAL AND VALUE ORIENTATIONS

A central assumption of this book is that certain political and value orientations influence how people perceive science and technology, which types and sources of information are appropriate, and how they perceive certain controversial policy issues such as those discussed in the forthcoming chapters. As Will Stelle with the National Marine Fisheries Service once commented concerning science and environmental issues, "Most people practice pick-and-choose ... agenda driven science in which the quality of the science is judged by the apparent results achieved. This is not biological science but political science" (cited in Blumm, 2002: 327). More specifically, in this book political ideology, postmaterialist values, and belief in positivistic science will be used to examine public attitudes and orientations toward climate change, immunizations, stem cell research, GMOs, and birth control/abstinence. Brief overviews of these concepts are provided below with a discussion of how they potentially affect controversial scientific issues.

Positivism: A positivistic scientific position holds that the scientific method and empirical data are critical in establishing scientific consensus by providing observable, replicable findings. In this view, science is the most reliable way to understand the world, and through strict methodological research can identify facts or truths (Steel et al., 2004) because, as Pigliucci and Boudry put it:

> Science (in the broad sense) is the practice that provides us with the most reliable (i.e., epistemically most warranted) statements that can be made, at the

time being, on subject matter covered by the community of knowledge, disciplines (i.e., on nature, ourselves as human beings, our societies, our physical constructions, and our thought constructions. (2013: 70)

Through a positivist lens, it is via science that we can gain deeper understanding about the world around us, which informs the policy process on issues regarding everything from human health to broader benefits to society and the environment.

In matters of public policy, a positivist view should lead to a public with high levels of scientific literacy (e.g., a greater understanding of the scientific process facilitates greater scientific literacy), which is assumed to be more conducive to bridging the science-policy gap. In this way, "increased communication and awareness about scientific issues will move public consensus toward the scientific consensus and reduce political polarization around science-based policy" (Hart and Nisbet, 2011: 1–2). Yet this focus on the science-knowledge deficit may not fully explain the gap between scientists and the public on controversial issues in public policy.

As previously discussed, people may rely on science that already complies with their beliefs. In this way, people are heuristic, focusing on values and ideological preferences to settle scientific "disputes" and choose a policy preference. Using longitudinal data from 1974 to 2010 of the General Social Survey, Gauchat explored trust in science among different groups. In this study, "conservatives were far more likely to define science as knowledge that should conform to common sense and religious tradition" (2012: 183), and conservatives "appear especially averse to regulatory science, defined . . . as the mutual dependence of organized science and government policy" (2012: 183). Thus trust in science, or more realistically, trust in scientific outcomes is subject to the same motivated reasoning that informs policy preferences.

A 2012 study found that ideological conservatives and moderates are "less confident in science than liberals" and that there was little difference among party lines between Democrats and Republicans (Gauchat, 2012: 177). Further, among conservatives, trust in science has declined since the 1970s, but has remained relatively static for liberals and moderates (Gauchat, 2012; Lewandowsky, Gignac, and Oberauer, 2013). In Chris Mooney's *The Republican Brain*, he attributes some of these trends as follows:

So it may be that greater openness and greater interest in learning about the world in all its complexity—not a general lack of motivated reasoning—bring liberals closer to science and facts. And it may *appear* [emphasis in the original] that conservatives are more motivated in their reasoning simply because, with policy preferences that are less likely to correspond to the kinds of knowledge that are acquired through curiosity and inquiry . . . conservatives simply have a more frequent need to resort to *political* [emphasis in the original] motivated reasoning to defend their beliefs. (2012: 259)

Ideology: In recent decades, there has been a growing polarization between liberals and conservatives regarding issues of science. In particular, Mooney (2005, 2012) suggests that the extent of this divide amounts to a partisan "war on science," with conservative Republicans outright denying the general scientific consensus on issues ranging from climate change to evolution. Recent studies appear to confirm this ideological divide. For example, a 2012 study found that ideological conservatives and moderates are "less confident in science than liberals" (Gauchat, 2012: 177). Further, among conservatives, trust in science has declined since the 1970s, but has remained relatively static for liberals and moderates (Gauchat, 2012; Lewandowsky, Gignac, and Oberauer, 2013). In 2014, a study by the Pew Research Center found that 42 percent of conservative Republicans and 30 percent of liberal Democrats perceive science as being liberal (Pew Research Center, 2015), which for conservatives may contribute to distrust in scientific findings, while for liberals it may have little to no effect. As Gauchat observed: "In the political sphere, the credibility of scientific knowledge is tied to cultural perceptions about its political neutrality and objectivity, which are crucial social resources for building consensus in ideologically polarized policy arenas" (2012: 168).

A further distinction concerns trust in science, specifically regarding the endorsement of free markets. On issues like climate change, "people who embrace a laissez-faire vision of the free market are less likely to accept that anthropogenic greenhouse gas emissions are warming the planet than people with an egalitarian-communitarian outlook" (Lewandowsky, Gignac, and Oberauer, 2013: e75637). Often, the endorsement of free markets is related to conservatism, with one study finding that both were related to the denial of climate science (Lewandowsky, Gignac, and Oberauer, 2013). Therefore, on many science issues pertaining to the environment, there are clear divisions between conservatives and liberals.

Conversely, on some issues, such as vaccinations, the liberal-conservative divide is not so evident. For example, Robert F. Kennedy Jr. is a vocal vaccine skeptic, lending credibility to the notion that liberals are more skeptical of (if not outright reject) vaccinations than conservatives. One exception, however, is the human papillomavirus (HPV) vaccine, which has met resistance from conservatives (Kahan et al., 2010); granted, this is due in part to the HPV vaccine's aim of immunizing teenagers before they become sexually active in order to prevent the passage of a sexually transmitted virus (Roll, 2007). Taken together, the picture becomes murky from an ideological viewpoint.

While conservatives tend to reject science related to climate change, stem cell research, and evolution, there is an argument that liberals reject science that pertains to issues like GMOs and vaccinations. Further, conservatives tend not to favor science conducted by government entities, while liberals are less likely to support research from for-profit organizations, especially those that challenge science to maintain or increase profits by undermining federal regulatory agencies (McCright and Dunlap, 2003). Interestingly, people on the political left are more likely to hold conspiratorial views (Lewandowsky, Gignac, and Oberauer, 2013), which could increase the likelihood of the rejection of science when large corporations (like Monsanto) are generating science in the support of policy.

Adding further complexity is the growing ideological polarization between Democrats and Republicans. Currently, there is far less ideological overlap between Democrats and Republicans than in previous decades. As the Pew Research Center found, "Today, 92% of Republicans are to the right of the median Democrat, and 94% of Democrats are to the left of the median Republican" (Pew Research Center, 2014). Unfortunately, the growing division in ideological positions is correlated to increased negative perceptions of those in the other party (Pew Research Center, 2014). The upside is that this division is most pronounced among those who are the most involved in politics and does not represent the majority of Americans, who do not uniformly hold either conservative or liberal views (Pew Research Center, 2014). However, it is most likely that the most ideological individuals who are engaged in the political process would either align or disagree with scientific consensus on issues.

Postmaterialism: Postindustrial nations, such as the United States, are that group of nations that enjoy relatively high levels of economic affluence, feature a high level of technological development, and manifest sustained political stability. The grouping of postindustrial nations includes Australia, Canada, Western Europe, New Zealand, Japan, and of course the West Coast of the United States, which will be the focus of the book's analyses. One characteristic that postindustrial nations have in common is that their historic economic development pattern has gone through each of the various phases of modernization. Their economies were at first characterized by highly labor-intensive agriculture and rural lifestyles; then by high output of manufactured goods, increasing urbanization, and strong economic growth; and finally by the predominance of service and tertiary sector activity. These countries share a history of enormous economic growth, most of which took place after the end of World War II. A decisive feature of this development was that, for the first time in history, the general population was able to partake in the substantial material benefits of this economic expansion; economic security and well-being were accorded to unprecedented proportions of the population in these countries.

Because of exceptional and rapid economic growth in the post–World War II period, and the increased importance of modern technology, the socioeconomic and political structures of these societies were altered, and social commentators began to take note of a new, novel stage of development (Dalton, 1988). This new stage or phase of socioeconomic development is commonly referred to as "postindustrial" or "advanced industrial" in the social science literature on modernity and postmodern intellectual thought (Inglehart, 1997).

Many studies have examined the psychological, social, political, and economic implications of postindustrialism (e.g., Huntington, 1974; Inglehart, 1997; Inglehart and Baker, 2000). The central theme of postindustrial change identified by these studies is the central role of technology. As Ronald Inglehart stated, "Technology is creating the Post-Industrial society just as it created Industrial society" (1977: 8). John Pierce and his colleagues argued that technology has led to the following characteristics of this new type of society:

> Postindustrial society and its attributes have had significant consequences for the nature of public policy challenges to be faced—such as environmental pollution, urban sprawl, biomedical discoveries for prolonging life—and in the kinds of policy claims made by activists among postindustrial citizens. The impact of postindustrial society on the public agenda of democratic politics has been manifested in an important way in the alteration of the distribution of value orientations among citizens in postindustrial countries. (1992: 9)

Other observers have also argued that nearly universal satisfaction of basic subsistence needs such as food and shelter has altered individual value structures among citizens, particularly younger cohorts. It is argued that these "higher order needs" have supplanted more fundamental subsistence needs as the motivation for much of societal behavior (Abramson and Inglehart, 1995; Inglehart, 1991; Flanagan, 1982;

Yankelovich, 1994). Value changes entailing greater attention to "post-materialist" or "postscarcity" needs are argued to have brought about changes in many types of personal behavior, including those related to the policy issues analyzed in this book.

However, postmaterialists also may be more aware of the negatives of scientific advancements (e.g., nuclear energy) and thus "have less confidence about what it can bring in the future, and be very wary of unintended consequences of scientific and technological development" (Price and Peterson, 2016: 67). A study by Price and Peterson found that "individuals holding post-materialist attitudes and living in countries with greater human and economic development (measured by higher internet access and tertiary enrollment, and lower infant mortality) have lower confidence in future-oriented science" (2016: 57). Somewhat perversely then, the more science and technology have benefited a society, the greater the potential for people to come to regard science with more skepticism.

METHODS AND APPROACH

This book seeks to understand people's views on science in general and the role science should play in the policy process concerning several controversial issues. We assume that research on motivated reasoning and cognitive bias is well established and relevant to the examination of value and ideological orientations and how they affect people's positions on scientific issues (e.g., Lodge and Taber, 2000; Slothuus and de Vreese, 2010). Therefore we are interested in seeing if these ideological biases remain consistent on highly contested social and environmental policy issues, specifically climate change, stem cell research, abstinence-only education, genetically modified food, and vaccinations. This book uses a variety of sources to examine these issues including reviews of policy advocate and interest group perspectives (both public pronouncements and written materials), and a 2016 survey of residents in Washington, Oregon, and California as a basis for ascertaining the preferred role of science in controversial policy issues and respondents' attitudes toward several of these controversial policy issues.

In order to investigate the relationship between values and perspectives on scientific controversies, each chapter examines the existing literature on each issue, provides an analysis of the arguments made by key actors in each policy domain, and examines data from a mail survey of sampled households that was conducted in California, Oregon, and Washington State during the winter and spring of 2016. Survey participants were

Table 1.1 Survey Response Bias

	U.S. Census (2015)	West Coast Survey
Mean age in years (18 and over)	48	49
Percent female (18 and over)	51	52
Percent associate's degree and higher	36	45

selected using a random sample provided by a national sampling company. Random address-based sampling from the U.S. Postal Service's Computerized Delivery Sequence file was used to generate 3,000 residential addresses. The file includes over 135 million residential addresses, which covers nearly all households in the United States. For the sample used here, the three West Coast states were sampled as a whole.

A modified version of Dillman's Tailored Design Method was used in questionnaire format with multiwave survey implementation (Dillman, 2007). Each contacted household was issued the following request for participation: "If available, we would prefer the person, 18 years old or older, who most recently celebrated a birthday to complete the survey." Three waves of first-class mail surveys were distributed (see Appendix A). Each mailing contained a copy of the survey, a hand-signed letter encouraging participation in the study, and a business postage prepaid envelope. Out of the 3,000 households contacted, we received responses from 1,316 participants (43.8% response rate). The response rate is calculated following the American Association for Public Opinion Research guidelines (2011). The demographic characteristics for respondents in comparison to U.S. Census rates for the region can be found in Table 1.1. The average age in years and the percent female are fairly representative of U.S. Census estimates for 2015, but the percent with an associate's degree or higher is 9 percentage points higher for survey respondents, which is typical for mail surveys.

The following section of analyses investigates the level of interest, subjective informedness, and preferred role for science and scientists in the policy-making process in general, followed by an examination of how values and ideology shape those attitudes using the 2016 public survey data.

ATTITUDES TOWARD SCIENCE AND SCIENTISTS

The West Coast public survey contained several introductory questions to ascertain how well informed respondents considered themselves concerning recent scientific discoveries and how often respondents talk science and

Table 1.2 Public Self-Assessed Awareness and Issue Salience of Scientific Issues

Question: How well informed do you consider yourself to be concerning recent scientific discoveries?

Percent	
19.5	Not informed
38.5	Somewhat informed
24.2	Informed
17.8	Very well informed

N = 1,485

Question: How often do you talk about issues related to science and scientific discoveries with your family, friends, or other acquaintances?

Percent	
10.2	Never
16.5	Hardly ever
40.0	Sometimes
33.3	Often

N = 1,483

scientific discoveries with family and friends. The data displayed in Table 1.2 provide some insight into these questions. For self-assessed level of informedness about "recent scientific discoveries," 19.5 percent said they were not informed, 38.5 percent considered themselves "somewhat informed," 24.2 percent considered themselves "informed," and 17.8 percent considered themselves "very well informed." Whether these results are seen positively or negatively in terms of scientific literacy is a judgment call, but with over 40 percent responding that they are "informed" and "very informed," it is a notable percentage.

In regard to how often respondents report discussing issues related to science and scientific discoveries, 10.2 percent responded "never," 16.5 percent replied "hardly ever," 40 percent said "sometimes," and a third of respondents replied "often." Once again, it is a judgment call whether these results are an indication of an engaged public in scientific discourse or an indication of indifference. However, with over 70 percent of respondents indicating they "sometimes" and "often" discuss issues related to science with friends or family members, it is also a sizeable portion of the public.

The next set of questions that were included in the survey concerns the proper role of science and scientists in the policy-making process.

These include what have been called "normal" or "traditional" roles for science and scientists (Steel et al., 2004). The role of scientists in this model is to provide relevant expertise about scientific findings that others in the policy-making process can use to formulate policies, and not to become advocates of particular policy preferences themselves. In this model, science is potentially perceived as having a special authority because of its independence and its power to "objectively" interpret the world. Also in this model, scientists can lose their credibility if they cross the line between science and policy advocacy (Lackey, 2007). We then get a "separatist" role for scientists; ideally, they are removed from policy making and decision making and serve as consultants or experts only. They are called upon as the need arises and as policy makers, managers, and the public require.

In recent years, a second model of science has emerged that challenges the normal, traditional role for science and scientists. The newly emerging model argues that science and scientists should become more integrated into policy-making processes. This emerging "integrative" model—also called "postnormal science"—calls for personal involvement by individual research scientists in public policy processes, providing expertise and even promoting specific strategies that they believe are supported by the available scientific knowledge (Ravetz, 1987; Steel and Weber, 2001).

The data displayed in Table 1.3 replicate a previous project conducted in the Pacific Northwest and nationally that asked the public, natural resource managers, and scientists their preferred role for science and scientists in the natural resource management process (Steel et al., 2004). The possible roles range from the normal science role described previously where scientists just report findings and leave others to make policy decisions, to more integrative roles where scientists help policy makers design policy, to a technocratic role where scientists make policy decisions themselves. While the five roles included in the table are not necessarily mutually exclusive, they do cover the full range of potential roles scientists could have in the policy process.

The roles least preferred by the public are on both ends of the possible continuum—the basic science or normal science role of just reporting scientific results and leaving others to make policy decisions, and the technocracy role of letting scientists make policy decisions. The strongest support is found for the integrative roles for scientists, with 66.2 percent agreeing that "scientists should report scientific results and then interpret the results for others in the policy process," and 68.6 percent agreeing that "scientists should work closely with policy makers and others to

Table 1.3 Public Preferences for the Role of Science and Scientists in the Policy Process

	Disagree *Percent*	Neutral *Percent*	Agree *Percent*
Scientists should only report scientific results and leave others to make policy decisions [N = 1,486].	46.9	20.2	32.9
Scientists should report scientific results and then interpret the results for others involved in policy making. [N = 1,485]	11.6	22.2	66.2
Scientists should work closely with managers and others to integrate scientific results in policy making. [N = 1,483]	9.2	22.2	68.6
Scientists should actively advocate for specific policies they prefer. [N = 1,485]	28.4	34.4	37.2
Scientists should make policy decisions. [N = 1,467]	52.5	34.6	12.9

integrate scientific results in policy making." So in the abstract, most of the public is fairly well supportive of scientists and science being integrated in policy making.

Survey respondents were also asked to select one of the scientific roles provided in Table 1.3 as "the best single description of your preferred role for scientists in the policy process." The results are displayed in Table 1.4 and indicate that "interpret results" (24.1%), "work to integrate results" (23.8%), and "advocate for specific policies" (27.9%) received the most support. Only 5.1 percent chose the technocracy response of "make policy decisions" and 19 percent chose the normal science response of "only report scientific results." These results do show public support for the integration of science and scientists into an abstract policy process, but how do different value types and ideology impact perceptions of science and scientists? That is the next topic to be covered.

IDEOLOGY AND VALUE ORIENTATIONS

The next set of analyses examines the impact of political ideology, postmaterialist values, and belief in positivism for the preferred role of science and scientists in the policy process. For each controversial scientific policy issue examined in this book, these three value and ideological orientations will be used to examine how they affect policy preferences and attitudes using the public survey data. The question used to assess political ideology in this study is: "On domestic policy issues, would you

Table 1.4 Preferred Role for Science and Scientists in the Policy Process

Question: Which is the best single description of your preferred role for scientists in the policy process (select only one from Table 1.2).

Percent	
19.0	Only report scientific results
24.1	Interpret results
23.8	Work to integrate results
27.9	Advocate for specific policies
5.1	Make policy decisions
N = 1,467	

consider yourself to be (please circle your response)," with a Likert-style scale of 1 = very liberal/left, 5 = moderate, and 9 = very conservative right. The distribution for the indicator is slightly skewed to the left as would be expected for the three "blue" states of California, Oregon, and Washington with large, very liberal cities (e.g., San Francisco, Portland, and Seattle; see Table 1.5). For the bivariate results presented in Table 1.5, the question was recoded into "liberal," "moderate," and "conservative" categories to simplify the presentation of results (i.e., 1 to 4 = liberal, 5 = moderate, and 6 to 9 = conservative).

Ronald Inglehart and his colleagues, who led the effort to design the World Values Survey, developed the indicator used to assess postmaterialist values. Inglehart and his colleagues have developed a "short form" measure of postmaterial values, which we implemented in this particular study. This measure asks respondents to choose two preferred value

Table 1.5 Value Orientation Indicators

Variable Name	*Variable Description*	*Results*
Ideology	*Political Ideology* 1 = Very liberal/left to 9 = Very conservative/right	Mean = 4.45 S.D. = 2.02 N = 1,479
Postmat	*Postmaterialist Values* 1 = Materialist, 2 = Mixed, 3 = Postmaterialist	Materialist = 4.5% Mixed = 58.3% Postmat = 37.2% N = 1,486
Positivism	*Positivism Additive Index* 6 = Weak Belief in Positivism to 30 = Strong Belief in Positivism	Mean = 19.92 S.D. = 5.23 N = 1,468

statements from a list of four available; having more say in government, protecting freedom of speech, fighting rising prices, and maintaining order were the choices.[1] Individuals choosing the first two are classified as postmaterialist, those picking the last two are categorized as materialist, and the other combinations are labeled "mixed." For the West Coast public survey, 37.2 percent of respondents can be classified as postmaterialists, 58.3 percent as mixed values, and only 4.5 percent as materialists.

To measure attitudes and beliefs about positivistic science, each respondent was asked his or her level of agreement or disagreement with a series of six statements that underlie many of the assumptions implicit in positivism, broadly construed.[2] These statements/assumptions came directly from the work of philosopher Karl Popper (1972) and have been previously developed and used to construct a "positivism index" for several NSF studies examining the role of science and scientists in the environmental policy process (Steel et al., 2004; Wolters et al., 2016). Agreement with the six statements can be interpreted as support for the important principles inherent in a positivistic perspective of science such as: "Science provides objective knowledge about the world," "Use of the scientific method is the only certain way to determine what is true or false about the world," and so on. The additive index has a range of 6 = weak belief in positivism to 30 = strong belief in positivism. The reliability coefficient (Cronbach's alpha) for the index is .774 and the mean score is 19.92. For the bivariate results presented in Table 1.6, the index was recoded into three categories of "high," "medium," and "low" levels of support for positivism (i.e., recoding is roughly the bottom one-third percent of respondents as "low support," the next roughly one-third percent as "medium support," and the top one-third percent for "high" level of support).

Regarding the impact of ideology on public preferences for science and scientists in the policy process, Table 1.6 presents bivariate analyses for each of the roles discussed previously in this chapter. For this table and many subsequent bivariate tables in this book, the Chi-square statistic (X^2) is provided for those readers interested in statistical significance. Chi-square is a bivariate statistical test applied to categorical data to determine how likely it is that any difference between an "expected" (null hypothesis) distribution and what was observed in the survey results arose by chance.

There are statistically significant results for all six scientific roles, with conservatives significantly more likely to support a normal/basic role for science and scientists (only report) and disagree with the technocratic role (make policy decisions) when compared to moderates and liberals. While liberals, moderates, and conservatives all strongly support an

Table 1.6 Ideology and the Role of Science and Scientists in Policy

		Liberal *Percent*	Moderate *Percent*	Conservative *Percent*
Only report scientific results.	Disagree	56.3	39.2	38.0
N = 1,479	Neutral	16.7	24.1	22.0
Chi-square = 47.53***	Agree	27.0	36.7	40.0
Interpret the results for others.	Disagree	11.8	11.0	12.0
N = 1,478	Neutral	17.7	25.1	27.6
Chi-square = 17.37**	Agree	70.5	63.8	60.4
Work to integrate results.	Disagree	8.1	6.9	13.6
N = 1,476	Neutral	17.6	29.5	22.7
Chi-square = 33.50***	Agree	74.3	63.6	63.7
Advocate for specific policies.	Disagree	20.9	32.6	37.8
N = 1,478	Neutral	31.8	38.5	34.8
Chi-square = 68.29***	Agree	47.3	29.0	27.3
Make policy decisions.	Disagree	44.8	52.6	66.4
N = 1,478	Neutral	39.6	35.4	24.3
Chi-square = 47.66***	Agree	15.5	12.1	9.3

*$p \leq .05$; **$p \leq .01$; ***$p \leq .001$

integrative role for scientists and science in the policy process, liberals were the most likely to with over 70 percent agreeing that scientists should interpret results and work to integrate results. In terms of advocacy, 47.3 percent of liberals agreed that scientists should advocate for specific policies compared to 29 percent of moderates and 27.3 percent of conservatives.

When asked to select a single preferred role for science and scientists in the policy process (Table 1.7), the favorite role for liberals is "advocate for specific policies" (37%), compared to moderates who preferred "interpret the results for others" (28.9%) and conservatives who also slightly preferred the "interpret" role over other roles (26.4%). The second choice for moderates and conservatives was the basic science role of just reporting results (25.8% and 25.4%, respectively). For liberals, their preferred second choice was "work to integrate results" (27.1%). Clearly these results confirm that liberals are more comfortable with prominent roles for science and scientists in the policy process when compared to moderates and conservatives. This is not to say that some conservatives are not supportive of science and scientists in the policy process, but that liberals are significantly more supportive of more prominent roles.

Table 1.7 Ideology and Preferred Role of Science and Scientists in Policy

	Liberal *Percent*	Moderate *Percent*	Conservative *Percent*
Only report scientific results.	11.6	25.8	25.4
Interpret the results for others.	19.9	28.9	26.4
Work to integrate results.	27.1	19.0	23.2
Advocate for specific policies.	37.0	18.8	21.2
Make policy decisions.	4.4	7.6	3.8

N = 1,460; Chi-square = 98.15; p = .000

Turning now to the impact of postmaterialist values on preferred roles for science and scientists in the policy process (Table 1.8), not surprisingly we find that postmaterialists were significantly more likely than other value types to disagree with the technocratic role (make policy decisions). They were also more likely to disagree with policy advocacy than other value types and were the least likely to disagree with the basic

Table 1.8 Postmaterialist Values and the Role of Science and Scientists in Policy

		Materialist *Percent*	Mixed *Percent*	Postmaterialist *Percent*
Only report scientific results.	Disagree	51.5	48.6	43.8
N = 1,486	Neutral	7.6	20.6	21.0
Chi-square = 10.82*	Agree	40.9	30.8	35.3
Interpret the results for others.	Disagree	31.8	8.9	13.4
N = 1,485	Neutral	18.2	22.6	22.1
Chi-square = 34.37***	Agree	50.0	68.5	64.6
Work to integrate results.	Disagree	6.1	9.6	9.0
N = 1,483	Neutral	12.1	22.1	23.5
Chi-square = 6.12	Agree	81.8	68.3	67.5
Advocate for specific policies.	Disagree	6.1	25.4	35.8
N = 1,485	Neutral	40.9	37.2	29.3
Chi-square = 37.19***	Agree	53.0	37.4	34.9
Make policy decisions.	Disagree	30.3	50.9	57.5
N = 1.485	Neutral	59.1	35.3	30.6
Chi-square = 24.57***	Agree	10.6	13.7	11.9

N = 1,467; Chi-square = 19.47; p = .013

Table 1.9 Postmaterialist Values and Preferred Role of Science and Scientists in Policy

	Materialist *Percent*	Mixed *Percent*	Postmaterialist *Percent*
Only report scientific results.	18.2	18.0	20.7
Interpret the results for others.	25.8	24.2	23.9
Work to integrate results.	15.2	22.7	26.6
Advocate for specific policies.	30.3	31.0	22.9
Make policy decisions.	10.6	4.2	5.9

N = 1,467; Chi-square = 19.47; $p = .013$

science role of just reporting results. All three value types strongly agreed with the integration role, ranging from 67.5 percent for postmaterialists, to 68.3 percent for mixed values and 81.8 percent for materialists. When examining the single preferred role for each value type (Table 1.9), postmaterialists are fairly spread out for the first four roles and had little support for technocracy (5.9%). Materialists and mixed-values respondents were more supportive for policy advocacy roles (30.3% and 31%, respectively) than postmaterialists (22.9%). Postmaterialists were slightly more likely to select an integrative role for science and scientists (26.6%) than mixed values (22.7%) and materialists (15.2%). In general, postmaterialists were found to be less supportive of prominent roles for science and scientists when compared to other value types, which was discussed earlier in the chapter. According to the literature, postmaterialists are more skeptical of experts and institutions and are more apt to engage in elite-challenging political activities when compared to other value types.

The final set of analyses conducted concerns beliefs about science and preferred roles for science and scientists in the policy process. More specifically, the belief in a positivistic science and the role of scientists in the policy process is examined. One would expect that people with high levels of belief in positivism would support more prominent roles for science and scientists in the policy process. When examining the results in Table 1.10, this is clearly the case. Those respondents with high levels of belief in a positivistic science are significantly less likely to support basic science roles (reporting results) and more likely to support the other five more prominent roles for science and scientists when

Table 1.10 Positivism Beliefs and the Role of Science and Scientists in Policy

		Low *Percent*	Medium *Percent*	High *Percent*
Only report scientific results.	Disagree	36.7	49.2	55.9
N = 1,468	Neutral	24.1	16.5	19.7
Chi-square = 44.03***	Agree	39.2	34.3	24.3
Interpret the results for others.	Disagree	13.6	11.4	8.8
N = 1,467	Neutral	29.5	21.3	15.6
Chi-square = 39.84***	Agree	56.8	67.4	75.6
Work to integrate results.	Disagree	11.9	10.5	5.1
N = 1,467	Neutral	32.2	22.5	11.0
Chi-square = 90.93	Agree	55.9	66.9	84.0
Advocate for specific policies.	Disagree	41.1	28.1	13.6
N = 1,467	Neutral	34.5	35.3	34.5
Chi-square = 115.10***	Agree	24.4	36.6	51.9
Make policy decisions.	Disagree	70.8	52.3	31.0
N = 1,467	Neutral	22.3	35.1	48.8
Chi-square = 157.41***	Agree	6.8	12.6	20.2

N = 1,467; Chi-square = 19.47; $p = .013$

compared to those respondents who have medium and low levels of belief in positivism. While majorities of all three levels of belief in positivism support the interpretation and integration roles, significantly higher levels of support were evident for high-level belief in positivism. In fact, 75.6 percent of high-level believers in positivism support the interpretation role and 86 percent support the integration role, compared to 55.9 percent to 66.9 percent of medium- and low-level believers. Over 20 percent of high-level believers even support a technocracy role for scientists!

Finally, when asked what the single preferred role for science and scientists should be in the policy process, only 7 percent of high-level believers supported a basic science role when compared to 18.7 percent of medium-level and 30.5 percent of low-level believers (Table 1.11). Not surprisingly, the most preferred role for high-level believers is policy advocacy (38.5%) compared to 26.8 percent for medium-level believers and 18.9 percent for low-level believers in positivism. Clearly belief in a positivistic science leads people to support more prominent roles for science and scientists in the policy process.

Table 1.11 Positivism Beliefs and Preferred Role of Science and Scientists in Policy

	Low *Percent*	Medium *Percent*	High *Percent*
Only report scientific results.	30.5	18.7	7.0
Interpret the results for others.	26.1	25.8	20.7
Work to integrate results.	21.2	24.5	26.2
Advocate for specific policies.	18.9	26.8	38.5
Make policy decisions.	3.3	4.2	7.5

N = 1,449; Chi-square = 118.72; $p = .000$

DISCUSSION AND BOOK PLAN

This introductory chapter has reviewed the growing controversy over science and policy and how values and ideology can influence people's acceptance of and/or opposition to the use of science in the policy process. Recent partisan squabbles over the science of global warming/climate change, the ethics of stem cell research, immunizations, and many other issues are indicative of a greater tendency for scientific research to get entangled in major ideological divisions in the political sphere. And this politicization of science is deepened by the role of government funding for scientific research and development, making government financial support a source of controversy that injects even more politics into scientific research policy. For example, Republicans in the U.S. Congress have been targeting geoscience and social science NSF funding in recent years, leading some observers to conclude that "many scientists believe those disciplinary distinctions are merely a ploy by Republicans to hide their real goal—curbing federally funded research on climate change and political science" (Mervis, 2015).

This book examines a variety of controversial environmental, health, and social science issues to identify how values and ideology are used to politicize and frame science and scientists in the policy process. More specifically, the approach of the book is to examine how political ideology, postmaterialist values, and belief in a positivistic science impact people's views on specific scientific policy issues. As the preliminary discussion and analysis of survey data indicate, values and ideology do have an impact on public preferences for the role of science and scientists in the policy process in the abstract. The remainder of the book will examine if these initial patterns are evident in the policy areas of climate change, vaccinations, GMOs, birth control/abstinence, and stem cell research.

NOTES

1. "There is a lot of talk these days about what our country's goals should be for the next ten or fifteen years. Listed below are some goals that different people say should be given top priority. Please mark the one you consider the most important in the long run. What would be your second choice? Please mark that second choice as well." [Response choices: having more say in government; protecting freedom of speech; fighting rising prices; and maintaining order.]

2. "Please indicate your level of agreement or disagreement with the following statements concerning the scientific process (1 = strongly disagree to 5 = strongly agree): Use of the scientific method is the only certain way to determine what is true or false about the world; The advance of knowledge is a linear process driven by key experiments; Science provides objective knowledge about the world; It is possible to eliminate values and value judgments from the interpretation of scientific data; Science provides universal laws or theories that can be verified; Facts describe true states of affairs about the world."

REFERENCES

Abramson, P., and R. Inglehart. *Value Change in Global Perspective.* Ann Arbor: University of Michigan Press, 1995.

American Association for Public Opinion Research. "Final Dispositions of Case Codes and Outcome Rates for Surveys." 2011.

Blumm, M. *Sacrificing Salmon: A Legal and Policy History of the Decline of Columbia River Salmon.* Portland, OR: BookWorld Publications, 2002.

Dalton, R. *Citizen Politics in Western Democracies: Public Opinion and Political Parties in the United States, Great Britain, West Germany, and France.* Chatham, NJ: Chatham House, 1988.

Dillman, D. A. *Mail and Internet Surveys: The Tailored Design Method.* 2nd ed. Hoboken, NJ: Wiley, 2007.

Flanagan, S. "Changing Values in Postindustrial Society." *Comparative Political Studies* 14 (1982): 99–128.

Gauchat, G. "Politicization of Science in the Public Sphere: A Study of Public Trust in the United States, 1974 to 2010." *American Sociological Review* 77, no. 2 (2012): 167–87.

Hart, P. S., and E. C. Nisbet. "Boomerang Effects in Science Communication: How Motivated Reasoning and Identity Cues Amplify Opinion Polarization about Climate Mitigation Policies." *Communication Research* (2011). doi: 10.1177/0093650211416646.

Huntington, S. "Postindustrial Politics: How Benign Will It Be?" *Comparative Politics* 6 (1974): 147–77.

Inglehart, R. *Culture Shift in Postindustrial Society*. Princeton, NJ: Princeton University Press, 1991.

Inglehart, R. *Modernization and Postmodernization: Cultural, Economic, and Political Change in 43 Societies*. Princeton, NJ: Princeton University Press, 1997.

Inglehart, R. *The Silent Revolution: Changing Values and Political Styles among Western Publics*. Princeton, NJ: Princeton University Press, 1977.

Inglehart, R., and W. Baker. "Modernization, Cultural Change, and the Persistence of Traditional Values." *American Sociological Review* 65 (2000): 19–51.

Janoff-Bulman, R. "To Provide or Protect: Motivational Bases of Political Liberalism and Conservatism." *Psychological Inquiry* 20 (2009): 120–28.

Jost, J. T. "The End of the End of Ideology." *American Psychologist* 61, no. 7 (2006): 651–70.

Jost, J. T., B. A. Nosek, and S. D. Gosling. "Ideology: Its Resurgence in Social, Personality, and Political Psychology." *Perspectives on Psychological Science* 3, no. 2 (2008): 126–36.

Kahan, D. M., D. Braman, G. L. Cohen, J. Gastil, and P. Slovic. "Who Fears the HPV Vaccine, Who Doesn't, and Why? An Experimental Study of the Mechanisms of Cultural Cognition." *Law and Human Behavior* 34, no. 6 (2010): 501–16.

Kahan, D. M., H. Jenkins-Smith, and D. Braman. "Cultural Cognition of Scientific Consensus." *Journal of Risk Research* 14, no. 2 (2011): 147–74.

Lackey, R. T. "Science, Scientists, and Policy Advocacy." *Conservation Biology* 21 (2007): 12–17.

Lewandowsky, S., G. E. Gignac, and K. Oberauer. "The Role of Conspiracist Ideation and Worldviews in Predicting Rejection of Science." *PLoS ONE* 8, no. 10 (2013): e75637.

Lodge, M., and C. Taber. "Three Steps toward a Theory of Motivated Political Reasoning." In *Elements of Reason: Cognition, Choice, and the Bounds of Rationality*, edited by A. Lupia, M. McCubbins, and S. Popkin, 183–213. Cambridge, MA: Cambridge University Press, 2000.

McCright, A. M., and R. E. Dunlap. "Defeating Kyoto: The Conservative Movement's Impact on U.S. Climate Change Policy." *Social Problems* 50, no. 3 (2003): 348–73.

Mervis, J. "Key House Republican Says 70% of NSF's Research Dollars Should Go to 'Core' Science—Not Geo or Social Science." *Science*, May 14, 2015.

Mooney, C. *The Republican Brain: The Science of Why They Deny Science—and Reality*. New York: Wiley, 2012.

Mooney, C. *The Republican War on Science*. Cambridge, MA: Basic Books, 2005.

National Science Foundation. "Science and Engineering Indicators 2016 (Nsb-2016-1)." January 2016.

Nir, L. "Motivated Reasoning and Public Opinion Perception." *Public Opinion Quarterly* 75, no. 3 (2011): 504–32.

Nyhan, B., and J. Reifler. "When Corrections Fail: The Persistence of Political Misperceptions." *Political Behavior* 32 (2010): 303–30.

Pew Research Center. "News Use across Social Media Platforms." May 26, 2016. http://www.journalism.org/2016/05/26/news-use-across-social-media-platforms-2016/.

Pew Research Center. "Political Polarization in the American Public: How Increasingly Ideological Uniformity and Partisan Antipathy Affect Politics, Compromise and Everyday Life." 2014.

Pew Research Center. "Public and Scientists' Views on Science and Society." January 29, 2015. http://www.pewinternet.org/2015/01/29/public-and-scientists-views-on-science-and-society/.

Pierce, J. C., Steger, M. A., Steel, B. S., and Lovrich, N. P. *Citizens, Political Communication, and Interest Groups: Environmental Organizations in Canada and the United States*. Westport, CT: Praeger, 1992.

Pigliucci, M., and M. Boudry. *Philosophy of Pseudo Science: Reconsidering the Demarcation Problem*. Chicago: University of Chicago Press, 2013.

Popper, K. R. *The Logic of Scientific Discovery*. New York: Routledge, 1972.

Porter, E. "Liberal Biases, Too, May Block Progress on Climate Change." *New York Times*, April 19, 2016.

Price, A. M., and L. P. Peterson. "Scientific Progress, Risk, and Development: Explaining Attitudes toward Science Cross-Nationally." *International Sociology* 31, no. 1 (2016): 57–80.

Ravetz, J. *The Merger of Knowledge with Power: Essays in Critical Science*. London: Mansell, 1990.

Ravetz, J. "Uncertainty, Ignorance, and Policy." In *Science for Public Policy*, edited by H. Brooks and C. Cooper. New York: Pergamon Press, 1987.

Redlawsk, D. P. "Hot Cognition or Cool Consideration? Testing the Effects of Motivated Reasoning on Political Decision Making." *Journal of Politics* 64, no. 4 (2002): 1021–44.

Roll, C. A. "The Human Papillomavirus Vaccine: Should It Be Mandatory or Voluntary?" *Journal of Health Care Law and Policy* 10 (2007): 421–30.

Slothuus, R., and C. H. de Vreese. "Political Parties, Motivated Reasoning, and Issue Framing Effects." *Journal of Politics* 72 (2010): 630–45.

Steel, B. S., P. List, D. Lach, and B. Shindler. "The Role of Scientists in the Environmental Policy Process: A Case Study from the American West." *Environmental Science and Policy* 7 (2004): 1–13.

Steel, B. S., and E. Weber. "Ecosystem Management, Decentralization, and Public Opinion." *Global Environmental Change* 11 (2001): 119–31.

Taber, C. S., D. Cann, and S. Kucsova. "The Motivated Processing of Political Arguments." *Political Behavior* 31, no. 2 (2009): 137–55.

Taber, C. S., and M. Lodge. "Motivated Skepticism in the Evaluation of Political Beliefs." *American Journal of Political Science* 50, no. 3 (2006): 755–69.

U.S. Nuclear Regulatory Commission. "Backgrounder on the Three Mile Island Accident." 2014. http://www.nrc.gov/reading-rm/doc-collections/fact-sheets/3mile-isle.html.

Westen, D., P. S. Blagov, K. Harenski, C. Kilts, and S. Hamann. "Neural Bases of Motivated Reasoning: An fMRI Study of Emotional Constraints on Partisan Political Judgment in the 2004 U.S. Presidential Election." *Journal of Cognitive Neuroscience* 18, no. 11 (2006): 1947–58.

Wolters, E. A., B. S. Steel, D. Lach, and D. Kloepfer. "What Is the Best Available Science: A Comparison of Marine Scientists, Managers, and Interest Groups." *Ocean and Coastal Management* 122 (2016): 95–122.

Yankelovich, D. "How Changes in the Economy Are Reshaping American Values." 1994. https://msu.edu/~mandrews/global/Changingvalues.pdf

CHAPTER 2

Genetically Modified Organisms (GMOs)

Any politician or scientist who tells you these [GMO] products are safe is either very stupid or lying.

—David Suzuki

INTRODUCTION

With the world's population rapidly increasing to 9 billion people by 2050 (United Nations, 2015), concern over meeting global food needs is rising. Most of the population growth is predicted for sub-Saharan Africa and the southeast parts of Asia. Concurrent with population growth, sub-Saharan Africa is expected to be one of the regions most impacted by climate change, resulting in less water, fewer acres of arable land, and overall less food security (Kotir, 2011). Concern over how to meet global food needs to combat food insecurity in places like sub-Saharan Africa requires addressing how to increase food production. Demand for high-yield crops and increased productivity suggest that a second Green Revolution[1] fueled by biotechnology, primarily genetically modified (GMO, or genetically modified organism) foods, may be necessary to meet current and growing food needs, as conventional agriculture requires more water and land, and is subject to diseases that GMO crops are resistant to.

Concurrent with the growing demand for increased crop yields globally, there is mounting concern in the United States over the safety of biotechnology. Although numerous studies have found that GMO foods[2] are safe for human consumption and the environment (DeFrancesco, 2013; European Commission, 2010) and produce higher crop yield and a reduction of pesticide use (Klümper and Qaim, 2014), recent opposition to GMO foods has signaled the public's concern over the use of biotechnology. As a result, some states, such as Vermont, have attempted labeling laws for foods that contain transgenic crops, that is, crops that have genetic material from another organism in them.[3] In addition, some farmers have switched back to non-GMO crops as consumers are demonstrating their preference by paying more for non-GMO foods. Further, companies like General Mills have announced they will no longer use transgenic foods in some of their products (Doering, 2015) in response to consumer preferences.

WHAT IS A GMO?

Recombinant DNA (rDNA) has been used in food production for the last several decades. This technology "allows selected individual genes to be transferred from one organism into another, also between nonrelated species" (World Health Organization [WHO], 2016). Biotechnology is unique in that it extracts the desired qualities in one species and adds that DNA to a different plant. The complicated aspect is that inserting the DNA alone will not necessarily produce the desired outcome. As such, "promoter" genes are inserted to activate the gene with the desired features. Agricultural biotechnology is different from traditional plant breeding in that it adds new genes to a plant. It is inherently this difference that separates supporters from opponents of agricultural biotechnology, specifically in GMO crops for human consumption.

As biotechnology gained momentum, so did concern over its use and safety. In 1975, research on biotechnology was ceased until the Asilomar Conference was held to discuss how the research should be conducted "with minimal risk to workers in laboratories, to the public at large, and to the animal and plant species sharing our ecosystems" (Berg et al., 1975: 1981). Essentially, the Asilomar Conference established a precautionary approach to biotechnology and outlined guidelines for future research. One of the organizers of the conference, Paul Berg, acknowledged that this technology "opened extraordinary avenues for genetics and could ultimately lead to exceptional opportunities in medicine, agriculture and industry," but that "the unfettered pursuit of these goals might have unforeseen and damaging consequences for human health and Earth's ecosystems" (Berg, 2008: 290).

With established biotechnology guidelines in place, agricultural biotechnology research progressed, and in the 1990s products came on the marketplace in the United States with the introduction of the FlavrSavr tomato and Monsanto's Bt (*Bacillus thuringiensis*) corn and cotton. The United States was not alone in doing this, as the European Union (EU) and several other countries also began producing GMO crops for consumers, particularly soybean, maize, and canola. In the mid-1990s, however, the United Kingdom suffered an outbreak of "mad cow disease." This crisis sparked growing concern in Europe about the safety of food (Paarlberg, 2013), and while completely unrelated to "mad cow disease," support for GMO foods diminished. Soon ships containing GMO crops were blockaded, consumers boycotted GMO products (Devos et al., 2008), and mandatory labeling of GMO foods in Europe went into effect in conjunction with strict restrictions (in some cases, bans) on GMO crops.

As the EU increased restrictions of GMO foods, the United States became more attentive to and concerned about biotechnology in food (much of which was already common in store-bought food, particularly products containing corn or soy, including animal products from animals fed on GMO crops). With a growing anti-GMO movement in the United States, four themes regarding public health and safety emerged: allergenicity, gene transfer, outcrossing, and human health.

Food Allergies: Although food allergies are increasing in the United States (Branum and Lukacs, 2008), it is unclear as to the reasons why (DeFrancesco, 2013). Because much of the noted rise in food allergies started in the 1990s, the same time GMO foods were becoming more prevalent in food, some anti-GMO activists have pointed to this correlation as causal. However, no link between transgenic foods and allergies has been demonstrated (DeFrancesco, 2013; WHO, 2016), although any new food introduced to consumers can elicit allergic reactions (conventional or GMO foods).

Antibiotic Resistance: Every year, roughly 2 million people in the United States are infected with bacteria, parasites, fungi, and viruses resistant to antibiotics, and several thousands of people die as a result (Centers for Disease Control and Prevention [CDC], 2016). If GMO foods with antibiotic-resistant genetic markers transferred cells into the human system, primarily through the gastrointestinal track or human genes, there is potential for increased human resistance to antibiotics and potentially negative impacts on human health (WHO, 2016). However, at this time, there is little to no evidence to suggest this has happened or will happen, and use of antibiotic gene transfers is discouraged by the WHO (2016).

Outcrossing: Agricultural land use, specifically seed dispersal, is not confined to clear borders, and therefore a potential for outcrossing, or "the migration of genes from GM into conventional crops," exists (WHO,

2016). Outcrossing has already occurred in low levels from feed crops for animals into crops intended for human consumption, and precautions are now being taken to limit contact between crops (WHO, 2016). However, the potential for outcrossing has rallied anti-GMO activists, who fear that non-GMO farmers can be sued for biotech seeds that naturally spread to their crops. In 1999, when a Canadian canola farmer was sued by Monsanto for illegally planting Roundup Ready canola seeds (the farmer claimed the seeds naturally dispersed to his fields), anti-GMO activists seized on this as evidence of the aggressive tactics of Monsanto to force farmers to buy their GMO products. However, Monsanto has proprietary ownership of the seeds (granted by their patent rights); therefore every farmer who plants Monsanto seeds gains rights to use the seeds but under the condition of not saving seeds for future use. In this case, the farmer lost when it was determined that he did, indeed, plant the seeds knowingly (and therefore violated Monsanto's patent), but the narrative of the farmer losing to aggressive business tactics continues to be cited by anti-GMO activists.

Public Health: Perhaps the primary concern about GMO foods pertains to public health. The United States is known for a precautionary approach to public health and safety, in large part because Congress and appointees of regulatory agencies are subject to public scrutiny and opinion (Lynch and Vogel, 2001). However, in terms of food production, the U.S. Food and Drug Administration evaluation of new food introduced to the market is subject to the "substantial equivalence" test, meaning if the food is substantially equivalent to food conventionally produced, then it is deemed as safe as the food currently on the market. This process is markedly different from the EU, which requires that new food be proven safe before becoming available to the public.

After several decades of research into transgenic crops, there is scant evidence to warrant the concern. Research has found that "compositional analyses of 129 transgenic crops submitted to the FDA for marketing authority from 1995 to 2012 have all failed to detect any significant difference . . . between the engineered plant and its nonengineered counterpart" (DeFrancesco, 2013: 795). In 2010, the European Commission released a study showing a decade of food safety related to GMO crops (European Commission, 2010). Further, in a study conducted by Italian scientists reviewing 10 years of GMO technology and 1,783 cases, each case failed to provide evidence of harm to human health (Nicolia et al., 2014).

Lack of evidence suggesting adverse human or ecological impacts from GMO crops, however, has not reduced anti-GMO sentiment. One glaring issue in each of these studies is the credibility of the scientists conducting the research. In 2009, 26 researchers studying corn-insect relationships sent an anonymous letter to the Environmental Protection Agency (EPA) asserting that biotech companies unduly hinder independent research (Pollack, 2009). Because of the aforementioned proprietary rights, biotech companies control who gets to research their products and in some cases "that permission is denied or the company insists on reviewing any findings before they can be published" (Pollack, 2009), which inherently undermines independent, unbiased research. It is this control over the technology that raises red flags about the veracity of claims suggesting no negative impacts from GMO crops. As author Meredith Niles notes, "The USDA does not conduct its own tests on biotech crop varieties when deregulating and approving them for planting in the United States. Instead, it relies on industry studies and data to access their safety on the environment and human health" (Niles, 2009). If, as asserted, research on biotech crops is overseen to some degree by biotech companies, then U.S. Department of Agriculture approval of GMO foods is based, essentially, on self-reported assessments by biotech companies.

In addition, studies that are critical of GMO crops are subject to intense scrutiny and even condemnations by fellow researchers and biotech companies. For example, a 2012 study published in *Food and Chemical Toxicity* claimed that rats that were fed Monsanto's GMO corn suffered adverse health effects, including tumor growth. In 2013, after several scientists attacked the methodology and scientific credibility of the study, the journal retracted it (Genetic Literacy Project, 2014). This was not an isolated attack on research perceived to undermine the benefits of biotechnology crops. Some researchers, such as Emma Rosi-Marshall and Jennifer Tank, who published preliminary findings suggesting that Bt maize impaired the development and growth of caddis-fly larvae and increased the death rate of adult caddis flies twofold, met with open hostility by other researchers who question the veracity and methodology of their findings (Waltz, 2009a). These "strong arm" tactics by scientific colleagues continue to solidify the ongoing distrust of biotech research and biotech companies like Monsanto. Further, Waltz suggests that these open hostilities hasten distrust, writing, "What is clear is that the seed industry is perceived as highly secretive and reluctant to share its products with scientists. This is fueling the view that companies have something to hide" (Waltz, 2009b). Much like the case of the Canadian canola farmer, studies that suggest potentially negative ecological impacts from biotechnology hold merit among anti-GMO activists who feel that similar health impacts can affect humans consuming GMO food, but that any ill effects of GMO foods are being "covered

The Golden Rice Controversy

The WHO estimates that in parts of Africa and Southeast Asia, from "250,000 to 500,000 vitamin A-deficient children become blind every years, half of them dying within 12 months of losing their sight" (WHO, 2016). In many of these regions where vitamin A deficiency occurs, pregnant mothers and young children lack foods or supplements with vitamin A. In 1999, scientists discovered that using genes that produce beta-carotene, a precursor to vitamin A, from a daffodil plant into a rice plant created what is known as "Golden Rice" (Paarlberg, 2013), which could be grown in developing countries to combat vitamin A deficiency. Over the years, Golden Rice has evolved from other transgenic combinations to provide even greater levels of vitamin A and would be provided to farmers free of charge to serve greater humanitarian goals. However, due to continued opposition from organizations like Greenpeace, Golden Rice has yet to be distributed to developing countries. In 2016, more than 100 Nobel Laureates posted an open letter to Greenpeace, the United Nations, and world governments to support GMO foods, particularly Golden Rice, for its ability to reduce blindness and death related to vitamin A deficiency.

up" by biotech companies, who essentially control all research pertaining to their products.

In addition to the aforementioned concerns, further apprehension pertaining to loss of biodiversity, terminator seeds (essentially a sterile seed, once planted cannot produce fertile seeds for future use), and increased pesticide use (to combat so-called superbugs and superweeds) has proven largely unwarranted (Paarlberg, 2013), yet continues to dominate the narrative of anti-GMO arguments. However, one overarching factor is the fear of a handful of corporations facilitating and profiting from biotechnology.[4] Through a narrative of unsafe foods, predatory lawsuits, monocrops, terminator seeds, and superbugs, the emphasis on the dangers and fear of these "Frankenfoods" coalesces the anti-GMO movement against "unnatural" food. Frankenfood "is a strong metaphor, which sidesteps rational arguments" (Devos et al., 2008), stirs strong emotions, and provides a clear image of an abnormal and freakish technological invention.

To be sure, there are questions about the benefits of GMO foods. Potential consumers of transgenic foods may not see direct benefits from this technology; thus with no perceived benefits, people focus on the potential costs. Advocates of biotechnology argue that benefits are evident by the reduction of pesticide use by 90 percent on GMO corn crops and the prevention of billions of dollars of crop losses with GM biotechnology (Folger, 2014). However, recent reports have found that the Bt crops, which are modified to have insecticidal toxins to reduce pest infestations and the use of conventional insecticides, no longer have the resistance as pests are developing resistance to Bt crops (Gassmann et al., 2014). Further, over the last 20 years crop yields have stagnated, perhaps coincidentally with reduced funding on biotechnology research (Dimick, 2014).

IDEOLOGY, VALUES, AND GMO BELIEFS

GMOs are inherently conflated with the relationship of the technology with those corporations that are profiting off of GMO science in food production. Innately, GMO foods are narratively constructed as "unnatural." Environmental activist Vandana Shiva's comment on GMOs captures this "unnatural" sentiment: "You cannot insert a gene you took from a bacteria to a seed and call it life. You haven't created life, instead you have polluted it" (Shiva, 2012). Conversely, the nexus between the technology and corporation may be the underlying problem. As Blancke (2015) argues:

> One may take issue with the involvement of multinationals or be concerned about herbicide resistance, but these issues have to do with how GM technology is sometimes applied and certainly to do not warrant resistance to the technology and to GMOs *in general* [emphasis in the original]. The emotional and intuitive basis of anti-GMO sentiments however prevents people from making these distinctions.

Many issues are involved in GMO food production, but the underlying controversy centers on protection of human and environmental health and profit-driven GMO food development. Do political ideology, positivism, and postmaterialism act as discerning mechanisms for pro- or anti-GMO beliefs? Intuitively, both a more liberal ideology and postmaterialist value system would potentially indicate anti-GMO sentiment, while a strong positivistic belief would support GMO food production.

Ideology: Popular media has held that opposition to biotechnology is the domain of ideological liberals (Kloor, 2012; Shermer, 2013; Fisher, 2013), perhaps in large part because vocal opposition comes from environmental groups, which are traditionally liberal in orientation, as well as from organic and local food advocates with a similar ideological support base. Further, recent GMO food labeling law propositions have come from traditionally liberal states, such as Vermont, California, and

Oregon, lending credibility to the claim that liberals are the primary opponents of GMO foods.

However, there is little conclusive evidence to support the claim that liberals are more likely to oppose biotechnology. Research regarding trust in science for information shows that liberals are more likely to trust science for information than their conservative counterparts (Gauchat, 2012; Hamilton, 2015), including information on biotechnology (Hamilton, 2015). Therefore liberals are arguably more supportive of public policy consistent with scientific consensus.

One challenge in breaking down the liberal anti-GMO narrative is the conflation of conspiracy beliefs and ideological worldviews. The conspiracist view on biotechnology paints GMO foods as the brainchild of large corporations who strive to control food supply. This plays well with the alternative argument that GMO foods are not safe for human consumption. Thus recent GMO labeling ballot initiatives, while coincidentally originate in states with a liberal population, may have less to do with political ideology and more to do with conspiracy beliefs.

A study examining the relationship between ideology, free-market worldviews, and conspiracist ideation did not find evidence for "the motivated rejection" of GMO foods based on politically liberal ideology (Lewandowsky, Gignac, and Oberauer, 2013: e75637). Instead, the research on biotechnology finds that opposition may stem more from the perception of the "current political and economic system as fair, legitimate, and stable" (Lewandowsky, Gignac, and Oberauer, 2013: e75637) and lack of familiarity or understanding of biotechnology (Hamilton, 2015).

Positivism: In some policy areas, like climate change, ideology stands out as a key indicator of climate change denial. However, support for biotechnology, specifically GMO foods, is more complicated in that there does not appear to be a significant partisan or ideological split among those opposed to GMO foods in some studies (Hiatt, 2015). In a recent Pew Research Center survey, 57 percent of Americans said that GMO foods are unsafe to eat, while only 37 percent said that GMO foods are safe to eat (Pew Research Center, 2015). This is in stark contrast to the 88 percent of scientists that felt it was safe to eat GMO food (Pew Research Center, 2015).

Perhaps tellingly, 67 percent of Americans said that scientists are not in agreement about health effects of GMO foods (Pew Research Center, 2015). Thus the lack of support over GMO technology may be lower because people are less familiar with what biotechnology is (Hamilton, 2015) and have a heightened sense of risk related to the unknown.

Further, a good deal of research on GMO foods emanates from large corporations, in which only 18 percent of Americans have "a great deal" of confidence (Gallup, 2016). To some degree lack of trust in scientific findings coming from big business (like Monsanto) is intuitive: the company directly benefits from supportive scientific findings. Thus there is a clear conflict-of-interest issue for agribusiness over biotechnology research that is not lost on the American public.

Postmaterialism: Two elements of the anti–GMO foods movement fall well into the postmaterial world of developed nations such as the United States—the precautionary principle and the notion of "natural" food. The precautionary principle follows Inglehart's (1990) assessment of postmaterial values because of the inherent focus on quality of life (e.g., human health) (Dunlap and McCright, 2011). And subsequently, it adheres to the precautionary approach the United Nations Conference on Environment and Development Principle 15 (1992) espouses:

> In order to protect the environment, the precautionary approach shall be widely applied by States according to their capabilities. Where there are threats of serious or irreversible damage, lack of full scientific certainty shall not be used as a reason for postponing cost-effective measures to prevent environmental degradation.

GMO advocates would point to the "lack of full scientific certainty" as a reason to reject crop biotechnology. From this perspective, regardless of potential benefits, the precautionary, postmaterial position would err on the side of caution, as transgenic crops have not *fully* demonstrated safety to human health.

Concurrent with the rise of biotechnology is a food movement that focuses on local, organic, and "slow food" produced on small farms (no agribusiness). In a postmaterial society, these food groups adhere to "food rules to express solidarity around secular values" (Paarlberg, 2013: 183). Thus the modern food movement goals are in part "to express through the diets we adopt a solidarity with others who share our identity, our values, or our particular life circumstances" (Paarlberg, 2013: 183). Of those values are foods produced "naturally," and often by small-scale or local farms. Disregarding whether it will be possible to feed the United States, let alone the world, with small-scale, organic methods, currently this type of food choice is restricted to all but a small cohort of people.

In addition, many in these modern food movements (including the anti-GMO movement) eschew agribusiness, especially corporations like Monsanto, who in 2014 was ranked the United States' third-most-hated company (Bennett, 2014). Many in the anti-GMO movement argue that these large businesses seek to control the food supply, primarily through monopolization of seeds and through patents (Savage, 2015). For postmaterialists, economic gain and quantity of production is less important than issues of social justice and human and environmental health. What is interesting is that only transgenic crops meet with opposition, considering that rDNA pharmaceuticals are uniformly accepted in the United States and Europe, perhaps because biomedical advances provide direct and obvious benefits to consumers (Herring, 2008).

ANALYSES

General Results: Three survey questions were used to ascertain respondent orientations toward GMO foods. Using a Likert scale, respondents were asked their level of disagreement or agreement with the following statements: (1) I purchase and/or eat genetically modified food; (2) I feel that GMO foods are not safe to consume; and (3) GMO crops reduce overall pesticide use. Results for these three questions are presented in Table 2.1. Overall, 42 percent agreed or strongly agreed that they purchase and/or eat genetically modified foods while 34.3 percent disagreed or strongly disagreed with this statement. When asked if GMO foods are not safe to consume, 36.9 percent agreed or strongly agreed with this statement, 31.2 percent disagreed or strongly disagreed, and 31.1 percent responded neutral. Finally, concerning the claim by GMO proponents that GMO crops reduce pesticide use, 32.6 percent agreed or strongly agreed that was the case, compared to 29.7 percent disagreeing or strongly disagreeing. The plurality of respondents responded "neutral" for this statement (39%), indicating uncertainty about the claim. Overall, respondents appear to be fairly divided in their orientations and use of GMOs in our three case study states. However, there appears to an interesting pattern in the data with many respondents indicating they purchase and/or eat GMOs, but at the same time indicating some uncertainty of the safety of such products.

The data presented in Table 2.2 provide correlations between the three GMO questions to examine consistency of responses. Correlation is a statistic that describes the degree of relationship between two rank-ordered variables and statistical significance. The correlation statistic ranges from a –1.0 (perfect inverse relationship) to1.0 (perfect positive relationship). The results indicate that all of the correlations are statistically significant at .000, with a range of strengths of relationships. Not surprisingly, the relationship between purchasing and/or eating GMOs is negatively

Table 2.1 West Coast Public and GMO Orientations

Question: How likely are you to agree or disgree with the following statements about genetically modified organisms (GMOs) and food?

	Strongly Disagree *Percent*	Disagree *Percent*	Neutral *Percent*	Agree *Percent*	Strongly Agree *Percent*
a. I purchase and/or eat genetically modified food.	18.6 N = 277	15.7 N = 233	3.1 N = 343	21.4 N = 318	20.6 N = 306
b. I feel that GMO foods are not safe to consume.	15.3 N = 227	15.9 N = 237	31.1 N = 462	17.6 N = 262	19.3 N = 287
c. GMO crops reduce overall pesticide use.	15.7 N = 234	12.2 N = 181	39.0 N = 580	17.0 N = 252	14.6 N = 217

correlated with a belief that GMOs are not safe to eat. However, the strength of the relationship is perhaps not as strong as one might assume at –.480. Similarly, there is a positive relationship between those who purchase and/or consume GMOs with a belief that GMOs reduce pesticide use (.282), which has been a rationale by GMO producers for GMO production and consumption. The final correlation presented indicates that those respondents who believe GMOs are not safe to use also disagree that GMOs reduce pesticide use (and vice versa), another relationship that one might expect given the discourse presented above concerning the proponents and opponents of GMO production and use (Tau b –.380).

Value Orientations: The next set of analyses examines the impact of political ideology, postmaterialist values, and belief in positivism for the

Table 2.2 Correlation Coefficients for GMO Orientations (Kendall's Tau b)

	a. I purchase and/or eat genetically modified food.	b. I feel that GMO foods are not safe to consume.
b. I feel that GMO foods are not safe to consume.	–.480*** N = 1,473	
c. GMO crops reduce overall pesticide use.	.282*** N = 1,463	–.380*** N = 1,463

****p* > .000

three GMO orientation questions. What can be said about the impact of political ideology on our three GMO orientation questions? The data displayed in Table 2.3 indicate that ideology does indeed affect how GMOs are viewed and is statistically significant at .000 (Chi-square = 32.174). The majority of conservatives (53.3%) agreed that they purchase and/or eat GMOs, compared to 38.4 percent of liberals and 38.4 percent of moderates. Similarly, 39 percent of liberals disagreed that they eat and/or purchase GMOs compared to 33.2 percent of moderates and 27.6 percent of conservatives.

Table 2.3 Political Ideology and GMO Orientations

a. I purchase and/or eat genetically modified food.

	Liberal *Percent*	Moderate *Percent*	Conservative *Percent*
Disagree	39.0	33.2	27.6
Neutral	22.6	28.4	19.1
Agree	38.4	38.4	53.3
N =	685	388	398
Chi-square =	32.174***		

b. I feel that GMO foods are not safe to consume.

	Liberal *Percent*	Moderate *Percent*	Conservative *Percent*
Disagree	28.7	27.2	40.9
Neutral	25.9	37.0	34.8
Agree	45.4	35.8	24.2
N =	687	386	396
Chi-square =	57.730***		

c. GMO crops reduce overall pesticide use.

	Liberal *Percent*	Moderate *Percent*	Conservative *Percent*
Disagree	37.2	22.0	18.9
Neutral	39.4	40.7	38.9
Agree	23.5	37.3	42.2
N =	681	381	396
Chi-square =	67.932***		

***$p > .000$

The other two GMO orientation questions also exhibit a similar pattern to the first question with liberals more negative and conservatives more positive toward GMOs. For conservatives, 40.9 percent disagreed with the statement that GMOs are not safe to eat compared to 28.7 percent of liberals and 27.2 percent of moderates. Over 45 percent of liberals agreed that GMOs are not safe to eat compared to 24.2 percent of conservatives and 35.8 percent of moderates. Once again, these results are statistically significant at .000 (Chi-square = 57.730).

For the question concerning the diminished use of pesticides with GMOs, 42.2 percent of conservatives agreed with the statement compared to 23.5 percent of liberals and 37.3 percent of moderates. Only 18.9 percent of conservatives disagreed that GMOs reduce pesticide use compared to 22 percent of moderates and 37.2 percent of liberals (Chi-square significant at .000). It does appear, however, that there is a great amount of uncertainty with this question with all three ideological groups either approaching or passing 40 percent indicating "neutral" responses.

Next we examine the impact of postmaterialist values on our three GMO orientation questions. Similar to the results for political ideology, we present bivariate results in Table 2.4. As was argued in the literature review above, postmaterialists should be less enamored with GMOs than those respondents with mixed or materialist values. Over 56 percent of respondents with postmaterialist values disagreed that they purchase and/or consume GMO products compared to 28.1 percent of materialists and 41.9 percent with mixed values. Over 16 percent of postmaterialists agreed that they purchase and/or eat GMOs compared to 48.4 percent of materialists and 35.8 percent of mixed-values respondents. As with the impact of political ideology on GMO orientations, the Chi-square results indicate a statistically significant relationship at .000 (Chi-square = 51.799).

When it comes to the safety of eating GMOs, 48.4 percent of postmaterialists agreed that GMOs are not safe to eat when compared to 47.2 percent of mixed-values respondents and 30 percent of materialists. However, there is an interesting pattern in these results with roughly a third of materialists responding to each of the categories—agree, disagree, and neutral. When examining the final GMO question concerning reduced use of pesticides with GMOs, there are even more pronounced differences. Postmaterialists are significantly more likely to disagree with the statement (42.0%) that GMOs reduce pesticides when compared to respondents with mixed values (20.9%) and materialist values (10.9%). On the flip side of the question, 26.3 percent of postmaterialists agreed that GMOs lead to reduced pesticide use while 56.3 percent of materialists agreed with the statement.

Table 2.4 Postmaterialist Values and GMO Orientations

a. I purchase and/or eat genetically modified food.

	Postmaterialist *Percent*	Mixed *Percent*	Materialist *Percent*
Disagree	56.1	41.9	28.1
Neutral	27.3	22.3	23.5
Agree	16.7	35.8	48.4
N =	551	860	66
Chi-square =	51.799***		

b. I feel that GMO foods are not safe to consume.

	Postmaterialist *Percent*	Mixed *Percent*	Materialist *Percent*
Disagree	26.6	26.9	34.7
Neutral	25.0	25.9	35.3
Agree	48.4	47.2	30.0
N =	553	858	64
Chi-square =	46.571***		

c. GMO crops reduce overall pesticide use.

	Postmaterialist *Percent*	Mixed *Percent*	Materialist *Percent*
Disagree	42.0	20.9	10.9
Neutral	31.6	43.5	32.8
Agree	26.3	35.6	56.3
N =	547	853	64
Chi-square =	85.453***		

***$p > .000$

The final set of bivariate analyses examines the impact of belief in positivism on GMO orientations (see Table 2.5). For the question concerning the purchase and potential consumption of GMO foods, those respondents with low levels of support for positivistic assumptions about science were the most likely to disagree with the statement (44.3%), while those with medium (28.3%) and high (31%) levels of support were less likely to disagree. Interestingly, there is less variation between the levels of support for positivism and those who agree with the statement with the low category at 39.4 percent and the high category at 42 percent.

Table 2.5 Belief in Positivism and GMO Orientations

a. I purchase and/or eat genetically modified food.

	Low *Percent*	Medium *Percent*	High *Percent*
Disagree	44.3	28.3	31.0
Neutral	16.3	27.4	27.0
Agree	39.4	44.3	42.0
N =	526	481	455
Chi-square =	39.979***		

b. I feel that GMO foods are not safe to consume.

	Low *Percent*	Medium *Percent*	High *Percent*
Disagree	30.9	31.9	30.9
Neutral	24.8	32.3	38.9
Agree	44.3	35.8	30.2
N =	524	483	453
Chi-square =	28.555***		

c. GMO crops reduce overall pesticide use.

	Low *Percent*	Medium *Percent*	High *Percent*
Disagree	47.1	21.9	14.1
Neutral	26.8	48.7	45.6
Agree	26.1	29.3	40.3
N =	522	474	454
Chi-square =	155.696***		

***$p > .000$

Concerning the question about the safety of GMO foods for consumption, the most notable difference between low, medium, and high levels of support for positivism concerns agreement with the question. Those with low levels of support for positivism were the most likely to agree with the statement that GMO foods are not safe to consume (44.3%), while medium- (35.8% percent) and high-level (30.2%) supporters were less likely to agree. As with the first bivariate analysis in Table 2.5, the Chi-square statistic is significant at .000.

For the third and final question concerning the reduction of pesticide use with GMO crops, we find the same pattern as the first two GMO orientation questions. Over 47 percent of respondents with low positivism disagreed that GMO crops reduce pesticides compared to 21.9 percent for the medium level and 14.1 percent for the high level. The table also reports that 26.1 percent with low positivism agreed with this statement in contrast to 40.3 percent of high-positivism-level respondents in agreement.

These bivariate results indicate that ideology, postmaterial values, and belief in positivism do play a role in shaping people's orientations toward GMOs. However, do ideology, values, and positivistic beliefs have an independent effect while controlling for several demographic factors? To answer this question ordinary least squares models are used to test the independent effect of ideology, postmaterialist values, and belief in positivism for each of the three GMO orientation indicators (using the original Likert scale coding of 1 = strongly disagree to 5 = strongly agree). Ordinary least squares is a linear regression procedure that is used to predict the value of dependent variables (in this case, the three GMO orientation questions) given the values of various independent variables (i.e., ideology, postmaterialist values, and belief in positivism). The models also include measures of age,[5] gender,[6] and education.[7] The results for the regression models are presented in Table 2.6.

F-test results indicate that all three models are statistically significant (meaning the models are a good fit), with adjusted R^2 scores ranging from .079 for the purchase/eat GMO foods question to .165 for the GMO reduces pesticides question (adjusted R^2 scores indicate the percentage of variation predicted in the dependent variables by the cumulative effect of the six independent variables). As with the previous bivariate analyses in Tables 2.4, 2.5, and 2.6, all three ideology and value orientation indicators have a statistically significant impact for all three dependent variables in the direction hypothesized, while controlling for demographic characteristics. Liberals are significantly less likely than conservatives to purchase and/or eat GMO foods, more likely to believe GMO foods are not safe to consume, and less likely to believe GMOs reduce pesticide use.

Those respondents with postmaterialist values, when compared to mixed-values and materialist-values respondents, also were significantly less likely to purchase and/or eat GMO foods, more likely to believe GMO foods are not safe to consume, and less likely to agree with the statement "GMO crops reduce overall pesticide use." The same pattern holds for beliefs in positivism with high levels of belief in positivism associated with more pro-GMO orientations such as purchasing and/eating

Table 2.6 Regression Estimates for GMO Orientations

Variable:	Purchase/Eat GMO Coefficient (Std. Error)	GMO Not Safe Coefficient (Std. Error)	GMO Reduce Pesticides Coefficient (Std. Error)
Age	–.005* (.002)	–.006** (.006)	.002 (.002)
Gender [1 = female; 0 = male]	–.571*** (.076)	.490*** (.173)	–.364*** (.292)
Education	. 053* (.024)	–.095*** (.022)	–.007 (.020)
Ideology	.085*** (.019)	–.090*** (.018)	.095*** (.016)
Postmat [1 = postmaterial values; 0 = else]	–.234** (.077)	.243*** (.071)	–.241*** (.065)
Positivism	.029*** (.007)	–.013* (.007)	.067*** (.006)
F-test =	21.742***	27.831***	48.109***
Adj. R^2 =	.079	.100	.165
N =	1,439	1,438	1,429

*$p < .05$; **$p < .01$; ***$p < .001$

GMO foods, believing GMO foods are safe to consume, and believing that GMOs reduce pesticide use. Those respondents with lower levels of belief exhibit more negative orientations to the GMO questions.

Finally, while the purpose of the ordinary least squares models was to investigate the independent effects of our ideology and value indicators on GMO orientations while controlling for several demographic variables, there are some statistically significant effects for the demographic variables. For example, older respondents were significantly more likely than younger respondents to purchase and/or eat GMO foods and to believe that GMO foods are safe to eat. Women were significantly more likely than men to not purchase and/or eat GMO foods, to believe that GMO foods are not safe to eat, and to believe that pesticide use is not reduced with GMOs. Finally, those respondents with higher education levels were more likely to agree that they purchase and/or eat GMO foods and to believe GMO foods are safe to eat.

DISCUSSION

According to the Pew Research Center (2015), a majority of Americans feel that GMO foods are unsafe to eat and believe that scientists are not in agreement about the health effects from eating GMO foods. All three states that currently have GMO labeling laws (Vermont, Connecticut, and Maine) are blue states (having voted for a Democrat for president in the last several presidential elections), perhaps signifying a liberal bias against GMO foods. Yet prior research indicates that liberals and Democrats tend to be more trusting of science than their conservative and Republican counterparts (Gauchat, 2012; Hamilton, 2015). Although to some degree technological advances would seem to enhance quality of life, a recent study found that postmaterialists had less confidence in technology and were more fearful of risks (Price and Peterson, 2016). Further, research demonstrates that overall, Americans are supportive of science (Pew Research Center, 2015), indicating that scientific consensus would hold weight on personal opinions regarding policy issues. Nonetheless, opposition to GMO foods is strong in the United States, indicating that there are people who, contrary to scientific consensus, do not understand or trust the science supporting GMO food safety (and biotechnology more broadly). Our analysis of West Coast state public opinion indicates that opposition is the domain of liberal, postmaterialist, and antipositivistic individuals.

With liberal support of science consistent over the last several years (Gauchat, 2012), it is curious that liberals lack support for GMO technology. One theory is that rejection of GMO technology is due to lack of trust in the institutions conducting research, such as biotech companies like Monsanto. Concern over control of food supply from big business is a theme that seems to rise to the surface when examining the lack of support on this issue. Only 41 percent of Democrats feel that government regulation of business is more harmful than beneficial to society, compared to 76 percent of Republicans (Pew Research Center, 2012). With support of the free market a cornerstone of conservative and Republican platforms, it would follow that liberals would demonstrate greater scrutiny of big business and have less trust in the science pertaining to the health and safety of GMO foods.

The liberal tendency toward anti-GMO attitudes is consistent with more liberal-leaning, postmaterialist values. With postmaterialists placing a higher value on environmental and human health, and expressing more suspicion of big corporations (Inglehart, 1997), the findings that both liberals and postmaterialists express GMO food skepticism is consistent with the tenets of liberalism and postmaterialism. However, it is these beliefs that have also invited criticism from those with anti-GMO attitudes as being somewhat antiquated. As Hiatt argues (2015), "The anti-GM movement seems to be fueled by a combination of anti-corporate suspicion, small-farm nostalgia and anxiety about unfamiliar technologies. It raises questions of environmental safety and corporate control as well as food safety."

The issue of GMO technology poses challenges to the ideation of liberals and postmaterialists. While both groups lean toward support of environmental and human health issues (arguably a humanitarian outlook as well), the lack of support for GMO technology fails to recognize the potential that this technology has for feeding the world's growing population. If GMO technology is to be one of the myriad of agricultural tools needed to produce enough food for the world's population, it would seem that perhaps a greater understanding of the technology and the scientific consensus on health and safety might sway opponents to reconsider their position. As researchers found, "The self-expressed willingness of those on the Left to defer to scientists indicates that political arguments based on objective, scientific research might have a powerful influence on opinion" (Blank and Shaw, 2015: 32). And it would further suggest that research on GMO food and crops needs to come from trusted sources, not from multinationals with a vested interest.

Finally, the belief in the scientific process to conduct robust, verifiable research has a positive effect on support for GMO technology. As expected, positivists expressed less hesitation to purchase and consume GMO foods, expressed less apprehension about the safety of the food, and showed a greater understanding of the benefits of GMO crops using less pesticide. The complicated issue with GMO foods, unlike support among liberals, postmaterialists, and positivists on other issues like climate change, is the need to disentangle worldviews from acceptance of science. Many who oppose GMO technology express a lack of trust in the institutions and organizations that produce much of the research supporting GMO technology. In one study, researchers found that "respondents do not trust many of the organizations that have the greatest resources and responsibilities for ensuring the safety of GM food" (Lang and Hallman, 2005: 1249). Therefore simply showing more research that substantiates the safety of GMO foods may not be sufficient to counter the pervasive anti-GMO rhetoric coming from what might be considered more trustworthy sources (especially from groups advocating for the public good).

Certainly independent research on biotechnology is necessary but not sufficient in mitigating concern over GMO crops as patent rights provide biotech companies with a degree of control over all research on their products. As previously stated, due to the proprietary rights of biotech companies over their products, "it's no secret that the seed industry has the power to shape the information available on biotech crops," especially considering that "anyone wishing to buy transgenic seeds has to sign what's called a technology stewardship agreement that says, among many things, that the buyer cannot conduct research on the seed, not give it to someone else for research" (Waltz, 2009b: 880). In large part because of the proprietary rights of biotech, even independent research is met with skepticism, as acquiring the seeds for research requires the tactic approval from biotech companies.

Contrary to many other science-policy issues where there is strong support from liberals and postmaterialists, scientific consensus supporting GMO technology does not seem to translate to public support from liberals and postmaterialists. Of the issues that the Pew Research Center examined on public and scientists' opinions on science, the discrepancy over GMO safety was the most pronounced with a 51-percentage-point difference (Pew Research Center, 2015). To be sure, biotechnology is difficult to understand, and many people are most likely unaware that it is relatively ubiquitous in products we are already using, particularly pharmaceuticals, to which there is little to no opposition. Yet food production from GMO technology offers a new conundrum that highlights issues of trust and risk. In this way, those who might otherwise demonstrate support more consistent with scientific consensus express skepticism of GMO foods.

For those holding strong support for science, the acceptance of scientific consensus is less riddled with doubt. However, for those who generally trust science, it may not simply be a matter of producing more research demonstrating the safety of GMO foods, but require more pronounced efforts to quell concerns over corporate agriculture. As Lang and Hallman point out, "Thinking through the uncertainties surrounding GM food calls for an examination of not only the scientific, environmental, and moral issues, but also an examination of the trust inherent in the social and organizational underpinnings of the technology" (2005: 1250). Doubtless, this is a challenging task, but one that may become imperative if the ability to meet the food demands of future populations depends, in part, on the ability to incorporate GMO technology into the panoply of agricultural options.

NOTES

1. The Green Revolution occurred between the 1930s and 1960s. During this time, technology was introduced to increase crop productivity. Through use of pesticides, mechanization, and irrigation systems, crop yields increased.

2. Also referred to as transgenic food.

3. Vermont passed a law in 2014 requiring GMO foods to be labeled. However, President Obama signed S. 764, which established federal guidelines to label GMO foods, but it is known as the DARK (Denying Americans the Right to Know) Act since it is notably less stringent that Vermont's labeling laws.

4. At the time of this writing, recently unsealed court documents indicate that Monsanto had written a paper that was later credited to academics (in order to seem academic driven, not corporate sponsored) pertaining to the safety of Roundup, the company's flagship product (Hakim, 2017). These types of practices by companies like Monsanto continue to fuel the flame of distrust over their products and increase calls for greater investigation into potentially negative public health and environmental outcomes.

5. The question used in response categories for age was: What is your current age in years ____________?

6. The question used in response categories for gender was: Your gender? ☐ Female ☐ Male

7. The question used in response categories for education was: What is your level of formal education?

1. Less than high school (grades 1–8)	5. Two year associate college degree (e.g., AA)
2. Some high school (no diploma)	6. College degree (e.g., BA, BS, AB)
3. High school graduate	7. Some postgraduate schooling (no degree)
4. Some college, no degree	8. Postgraduate/professional degree (e.g., MA, JD)

Other? ________________________________

REFERENCES

Bennett, D. "Inside Monsanto, America's Third-Most-Hated Company." *Bloomberg*, June 4, 2014.

Berg, P. "Asilomar 1975: DNA Modification Secured." *Nature* 455 (2008): 290–91.

Berg, P., D. Baltimore, S. Brenner, R. O. Roblin III, and M. F. Singer. "Summary Statement of the Asilomar Conference on Recombinant DNA

Molecules." *Proceedings of the National Academy of Sciences* 72, no. 6 (1975): 1981–84.

Blancke, S. "Why People Oppose GMOs Even Though Science Says They Are Safe: Intuition Can Encourage Opinions That Are Contrary to the Facts." *Scientific American*, August 18, 2015.

Blank, J. M., and D. Shaw. "Does Partisanship Shape Attitudes toward Science and Public Policy? The Case for Ideology and Religion." *ANNALS of the American Academy of Political and Social Science* 658, no. 1 (2015): 18–35.

Branum, A. M., and S. L. Lukacs. "Food Allergy among U.S. Children: Trends in Prevalence and Hospitalizations." Centers for Disease Control and Prevention, 2008.

Centers for Disease Control and Prevention. "Antibiotic/Antimicrobial Resistance." 2017. https://www.cdc.gov/drugresistance/.

DeFrancesco, L. "How Safe Does Transgenic Food Need to Be?" *Nature Biotechnology* 31, no. 9 (2013): 794–802.

Devos, Y., P. Maeseele, D. Reheul, L. Van Speybroeck, and D. Waele. "Ethics in the Societal Debate on Genetically Modified Organisms: A (Re)Quest for Sense and Sensibility." *Journal of Agricultural and Environmental Ethics* 21, no. 1 (2008): 29–61.

Dimick, D. "Here's Why We Haven't Quite Figured Out How to Feed Billions More People." *National Geographic*, October 4, 2014.

Doering, C. "Farmers Turn to GMO-Free Crops to Boost Income." *Des Moines Register*, April 18, 2015.

Dunlap, R. E., and A. M. McCright. "Organized Climate Change Denial." In *The Oxford Handbook of Climate Change and Society*, edited by J. S. Dryzek, R. B. Norgaard, and D. Schlosberg, 99–115. Oxford: Oxford University Press, 2011.

European Commission. "A Decade of EU-Funded GMO Research." Brussels: European Commission Food, Agriculture and Fisheries and Biotechnology, 2010.

Fisher, M. "The Republican Party Isn't Really the Anti-science Party." *Atlantic*, November 11, 2013.

Folger, T. "The Next Green Revolution." *National Geographic*, October 2014.

Gallup. "Americans' Confidence in Institutions Stays Low." June 13, 2016.

Gassmann, A. J., J. L. Petzold-Maxwell, E. H. Clifton, M. W. Dunbar, A. M. Hoffmann, D. A. Ingber, and R. S. Keweshan. "Field-Evolved Resistance by Western Corn Rootworm to Multiple *Bacillus Thuringiensis* Toxins in Transgenic Maize." *Proceedings of the National Academy of Sciences* 111, no. 14 (2014): 5141–46.

Gauchat, G. "Politicization of Science in the Public Sphere: A Study of Public Trust in the United States, 1974 to 2010."*American Sociological Review* 77, no. 2 (2012): 167–87.

Genetic Literacy Project. "Scientists React to Republished Séralini GMO Maize Rat Study." June 24, 2014. https://www.geneticliteracyproject.org/2014/06/24/scientists-react-to-republished-seralini-maize-rat-study/.

Hakim, D. "Unsealed Documents Raise Questions on Monsanto Weed Killer." *New York Times*, March 14, 2017.

Hamilton, L. C. "Conservative and Liberal Views of Science: Does Trust Depend on Topic?" *Regional Issue Brief #45*, edited by Carsey School of Public Policy at the Scholars' Repository, 2015.

Herring, R. J. "Opposition to Transgenic Technologies: Ideology, Interests and Collective Action Frames." *Nature* 9 (2008): 458–63.

Hiatt, F. "Science That Is Hard to Swallow." *Washington Post*, February 8, 2015.

Inglehart, R. *Culture Shift in Advanced Industrial Society*. Princeton, NJ: Princeton University Press, 1990.

Inglehart, R. *Modernization and Postmodernization: Cultural, Economic, and Political Change in 43 Societies*. Princeton, NJ: Princeton University Press, 1997.

Kloor, K. "GMO Opponents Are the Climate Skeptics of the Left." *Slate*, September 26, 2012.

Klümper, W., and M. Qaim. "A Meta-analysis of the Impacts of Genetically Modified Crops." *PLoS ONE* 9, no. 11 (2014): e111629.

Kotir, J. "Climate Change and Variability in Sub-Saharan Africa: A Review of Current and Future Trends and Impacts on Agriculture and Food Security." *Environment Development and Sustainability* 13, no. 3 (2011): 587–605.

Lang, J. T., and W. K. Hallman. "Who Does the Public Trust? The Case of Genetically Modified Food in the United States." *Risk Analysis* 25, no. 5 (2005): 1241–52.

Lewandowsky, S., G. E. Gignac, and K. Oberauer. "The Role of Conspiracist Ideation and Worldviews in Predicting Rejection of Science." *PLoS ONE* 8, no. 10 (2013): e75637.

Lynch, D., and D. Vogel. "The Regulation of GMOs in Europe and the United States: A Case-Study of Contemporary European Regulatory Politics." Council on Foreign Relations. April 5, 2001.

Nicolia, A., A. Manzo, F. Veronesi, and D. Rosellini. "An Overview of the Last 10 Years of Genetically Engineered Crop Safety Research." *Critical Reviews in Biotechnology* 34, no. 1 (2014): 77–88.

Niles, M. "How Biotech Companies Control Research on GMO Crops." *Grist*, February 23, 2009.

Paarlberg, R. *Food Politics: What Everyone Needs to Know*. 2nd ed. New York: Oxford University Press, 2013.

Pew Research Center. "Deepening Divide between Republicans and Democrats over Business Regulation." 2012.

Pew Research Center. "Public and Scientists' Views on Science and Society." 2015.

Pollack, A. "Crop Scientists Say Biotechnology Seed Companies Are Thwarting Research." *New York Times*, February 19, 2009.

Price, A. M., and L. P. Peterson. "Scientific Progress, Risk, and Development: Explaining Attitudes toward Science Cross-Nationally." *International Sociology* 31, no. 1 (2016): 57–80.

Savage, S. "Who Controls the Food Supply?" *Forbes*, June 26, 2015.

Shermer, M. "The Liberals' War on Science: How Politics Distorts Science on Both Ends of the Spectrum." *Scientific American*, February 1, 2013. https://www.scientificamerican.com/article/the-liberals-war-on-science/.

Shiva, V. Interview by Bill Moyers. "Vandana Shiva on the Problem of Genetically Modified Seeds." July 13, 2012. http://billmoyers.com/segment/vandana-shiva-on-the-problem-with-genetically-modified-seeds/.

United Nations, Department of Economic and Social Affairs, Population Division. "World Population Prospects: The 2015 Revision, Key Findings and Advance Tables." 2015.

United Nations Conference on Environment and Development. "Report of the United Nations Conference on Environment and Development: Rio Declaration on Environment and Development." 1992.

Waltz, E. "Battlefield: Papers Suggesting That Biotech Crops Might Harm the Environment Attract a Hail of Abuse from Other Scientists." *Nature* 461, no. 3 (2009a): 27–32.

Waltz, E. "Under Wraps: Are the Crop Industry's Strong-Arm Tactics and Close-Fisted Attitude to Sharing Seeds Holding Back Independent Research and Undermining Public Acceptance of Transgenic Crops?" *Nature Biotechnology* 27, no. 10 (2009b): 880–82.

World Health Organization. "Frequently Asked Questions on Genetically Modified Foods." 2016.

CHAPTER 3

Childhood Vaccinations

Just the other day, 2 years old, 2½ years old, a child, a beautiful child went to have the vaccine, and came back, and a week later got a tremendous fever, got very, very sick, now is autistic.

—Donald J. Trump during a campaign debate, September 2015

INTRODUCTION

On June 30, 2015, California governor Jerry Brown signed into effect Senate Bill 277 and, with the swipe of a pen, effectively eliminated all religious and personal belief exemptions for vaccinations, making California the third state to eliminate these exemptions. For vaccine advocates, this move signified a science-based move to improve protection from vaccine preventable diseases (VPDs) that have increasingly been cropping up around the nation, a sentiment shared by Governor Brown when he stated at signing:

> The science is clear that vaccines dramatically protect children against a number of infectious and dangerous diseases . . . While it's true that no medical intervention is without risk, the evidence show that immunization powerfully benefits and protects the community. (Siders, 2015)

The governor's signing of the bill had the support of 67 percent of California public school parents, who believe that unvaccinated children should not attend public school (Baldassare et al., 2015). But it drew condemnation from antivaccination parents who felt the governor's decision forced them to choose between the health of their child and pulling their child from public school.

Brown's signing statement articulated concisely the vaccination debate. On one side are vaccine supporters who point to vaccination success for the eradication or minimization of several debilitating and deadly diseases. On the other side are vaccine opponents who point to limited, albeit terrifying, "evidence" of vaccine injury. Somewhat contrarily, the success of vaccines has minimized the impacts of disease, allowing focus to shift onto vaccine injuries. Paradoxically, due to the high efficacy of immunizations, these injuries now seem more prevalent than the preventable diseases themselves (André, 2003). However, in reality, severe vaccine reactions are extremely rare; fewer than 1 in 1 million people are affected (CDC, 2016b).

Vaccinations arose out of a need to protect human populations from debilitating and deadly diseases. Today, there are vaccines for 17 VPDs that have successfully reduced, or eradicated, the number of infections and deaths in the United States. A review of infections prevaccine and postvaccine demonstrates the success of some common vaccines (see Table 3.1).

Millions of lives have been saved and countless others have been spared from incapacitating diseases due to vaccinations. For example, smallpox, a disease that once killed 3 out of every 10 infected people and left those who survived with severe scarring, has been eradicated globally since 1980 (CDC, 2016a). The disease, thought to be inflicting people since the Egyptian Empire (CDC, 2016a), has killed millions of people,

Table 3.1 Comparison of Prevaccine Estimated Average Annual Cases vs. Most Recent Reported Vaccine Cases from 2006, United States

Vaccine-Preventable Disease	Prevaccine Estimated Annual Average Cases in United States	Most Recent Reported Postvaccine Cases (2006)	Percent Reduction Prevaccine Annual Estimate vs. Most Recent Reported
Diphtheria	21,053	0	100
Measles	530,217	55	99.9
Mumps	162,344	6584	95.9
Pertussis	200,752	15,632	92.2
Poliomyelitis, paralytic	16,316	0	100
Smallpox	29,005	0	100

Source: S. W. Roush and T. V. Murphy, "Historical Comparisons of Morbidity and Mortality for Vaccine-Preventable Diseases in the United States," JAMA 298, no. 18 (2007): 2155–63.

"hastened the decline of the Roman Empire," and led to the "collapse of the Aztec and Incan kingdoms" (Mnookin, 2011: 23). In addition, prior to the measles vaccination program in 1963, measles afflicted an estimated 3–4 million people each year in the United States, with the worst cases resulting in death (400–500 cases) (CDC, 2016b). Today, measles infections have been reduced by 99 percent due to vaccine efficacy (CDC, 2016b). Finally, pertussis (commonly known as whooping cough) once infected 200,000 children a year, resulting in the deaths of 9,000 of those children (CDC, 2017b). Since the pertussis vaccine, about 10,000 to 40,000 children are infected each year, with an average of 20 deaths per year (CDC, 2017b). Without vaccines, many of these diseases would still be prevalent in the United States, imposing huge tolls on public health.

Although vaccine programs have been very effective at reducing rates of infection, some VPDs are reemerging in the United States. In 2015, a measles outbreak in Disneyland infected over 125 people (CDC, 2015). Most of the infected people had either abstained from obtaining the vaccine or could not be vaccinated because they were too young (NBC News, 2015). The whooping cough epidemic in California in 2010 infected over 9,000 people, resulting in the hospitalization of 808 people and the death of 10 babies (CDC, 2014a). Both outbreaks have been blamed, in part, on antivaccinators.[1] In the Disneyland case, it is believed that an unvaccinated individual traveling or living abroad brought the virus into the United States (NBC News, 2015) and infected, almost exclusively, unvaccinated people.

At a time when U.S. vaccination rates remain high (CDC, 2014b) and vaccines are widely available, there persists a small group of people expressing criticism, skepticism, and rejection of vaccinations. The resurgence of these VPDs creates concern that antivaccinators provide a conduit for VPDs to spread throughout a community due to lower levels of immunization. The only way to contain or eradicate VPDs is through the immunization of at least 90 percent or more of a population (what is often referred to as "herd immunity"). Consequently, people who choose not to vaccinate put human populations at greater risk by introducing or spreading VPDs. Without vaccination compliance, more cases of VPDs will continue to emerge in pocket regions, and may have far-reaching impacts on human populations.

There are many reasons why parents may abstain from vaccinating their child, such as cost, ideological views, medical concerns, and religious restrictions. Cost of vaccination was once a primary deterrent to following a strict childhood vaccination schedule. However, the Vaccines for Children Program enacted under the Clinton administration established a federally funded program to provide immunizations at no cost to children who might not otherwise be vaccinated due to financial concerns.[2] Today the primary exemptions for antivaccinators are ideological (philosophical), religious, or medical.[3] While some religions strictly limit vaccinations due to the vaccine's preservation process (e.g., through porcine products), a meta study by Grabenstein found that in many cases, "ostensibly religious reasons to decline immunization actually reflected concerns about vaccine safety or personal beliefs . . . rather than theologically based objections per se" (2013: 2011).[4]

At present, almost all states still allow for religious exemptions, but only about 20 allow for philosophical exemptions. However, after the 2015 Disneyland measles outbreak, 19 states introduced legislation to tighten laws pertaining to religious or philosophical exemptions (Breslow and Amico, 2015). Because community protection from VPDs requires high levels of vaccine compliance, the more people who abstain for nonmedical reasons, the more communities and medically vulnerable kids who cannot get immunized are put at a higher risk for infection.

Vaccinations are not without some adverse reactions, such as pain, swelling, and fever. In very rare cases, there have been serious reactions reported, but generally it is very difficult to determine if the vaccine caused the reaction (André, 2003). This creates a complex problem for parents deciding whether or not to immunize their children. Parents must weigh the benefits of vaccination (protection for their child and community from VPDs) against the perceived harms their child could suffer by getting the vaccine. For some people, choosing to immunize their children may represent a risk they will not take, based on the perceived possibility of vaccines harming their children (even finding contracting the disease less harmful than potential vaccine "reactions").

THE CONTROVERSY

While vaccines have been the subject of safety concerns since their invention, a seminal turning point in the current antivaccination movement came in 1998 when findings by British doctor Andrew Wakefield suggested a link between the MMR vaccination and autism. Although Wakefield's study has never been replicated and is now considered erroneous (Wakefield's medical license has since been revoked due to his misleading results from the autism-MMR study), his "findings" and continued insistence that vaccines are harmful to children's health established seeds of doubt as to the safety and efficacy of vaccines, lending credibility to the antivaccination movement.

Refrigerator Mothers

The primary driver of current vaccine skepticism is the increasing rate of autism now being reported. In a study conducted in California, researchers noted a sevenfold increase from 1990 births to 2001 births that could not be explained by an increase in diagnosis of autism (Hertz-Picciotto and Delwiche, 2009). Antivaccinators suggest that it is either the contents of the vaccines or the increased number of vaccines for children that could explain this increase. However, current research now suggests that the increase in autism may have more to do with maternal exposure to air pollution, persistent organic pollutants (like industrial chemicals) and pesticides, in addition to maternal lifestyle choices like the intake of vitamin D or folic acid (Lyall, Schmidt, and Hertz-Picciotto, 2014).

Parents of autistic children seek to understand why their child is autistic. The suggestion that maternal lifestyle choices and exposure to harmful pollutants unintentionally brings up previous theories that the mothers behavior caused her childs autism. Leo Kanner, a psychiatrist at Johns Hopkins University who originated the diagnosis of autism, conducted a study in 1943 of several autistic children from educated families in an academic community (PBS, 2002). Kanner determined that the cold, intellectual nature of the parents, particularly the mothers, was what led to autism in children. This theory, later coined the refrigerator mother, lasted until the late 1960s when it was countered by another researcher, Bernard Rimland.

However, the suggestion by Kanner that autism was a result of the mothers behavior continues to resonate today. Parents, and mothers in particular, are caught in a conundrum about vaccination. Choosing to vaccinate implies to some that mothers are willing to risk their childs health and put their trust in Big Pharma to protect their child. If they chose not to vaccinate, they are sacrificing the protection of the whole community, particularly other children. Although vaccines demonstrably do not cause autism, parents of autistic children (and maybe more so mothers) must (however unwarranted) contend with choices they made that resulted in their autistic child. Psychologically, it makes sense. People seek answers to make sense of the world around them. For parents of autistic children, the desire to understand why their child is autistic may provide a barrier to historical attacks that put the blame on parents for their childs autism.

In the search to identify the purported culprit in vaccines that cause autism, the focus shifted onto thimerosal, a mercury-based preservative used in some vaccines to prevent contamination by bacteria and fungus (Christensen, 2017). Because thimerosal was once used in multidose vials of vaccines, concern arose over the cumulative effects of mercury on young children. Thimerosal contains ethylmercury, not methylmercury, which is found in certain fish species and can become toxic for people at high levels (hence the suggestion to reduce the consumption of fish known to contain methylmercury, particularly during pregnancy). Unlike methylmercury, which remains in the body for an extended period of time, ethylmercury is quickly evacuated from the human body and therefore does not accumulate over time. However, due to the vaccination schedule where routine vaccinations occur within the first six months to one year of life, as a precautionary measure thimerosal (and therefore ethylmercury) was removed from most childhood vaccines in 2001 to limit any cumulative effect of mercury.[5]

At this time, the major objections to vaccines are twofold: first, thimerosal-preserved vaccines, and second, the vaccination schedule itself, which many feel is too many too soon, allowing for the compounding effect of those vaccines to induce autism or other neurological conditions. In terms of thimerosal's connection to autism, there are no studies drawing conclusive evidence that thimerosal increases risk of autism. However, many studies do find there is *no* connection between thimerosal and autism (Hviid et al., 2003; Stehr-Green et al., 2003; Gerber and Offit, 2009; Price et al., 2010) and *no* connection between the MMR vaccine and autism (Madsen et al., 2002; Taylor et al., 2002). Further, studies do not find any evidence suggesting that the number of vaccines within the first one or two years of life is linked with the onset of autism (Smith and Woods, 2010; DeStefano, Price, and Weintraub, 2013). In a meta-analysis of studies involving over 1 million children examining possible links between vaccines and autism by Taylor, Swerdfeger, and Eslick in 2014, findings suggest that:

> Vaccinations are not associated with the development of autism or autism spectrum disorder. Furthermore, the components of the vaccines (thimerosal or mercury) or multiple vaccines (MMR) are not associated with the development of autism or autism spectrum disorder. (3623)

In short, there is abundant evidence finding *no* link between thimerosal and autism, multiple vaccines (like MMR) and autism, or a connection between the vaccine schedule (too many, too soon) and autism. Critical to understanding the antivaccination stance is the efficacy-versus-harm issue related to vaccines. The challenge is balancing a personal desire to protect children, and a social responsibility to protect the community at large.

Table 3.1 illustrates how postvaccine availability reduced infection by greater than 90 percent (in some cases 100 percent) for several once common diseases. The postvaccine cases outlined in Table 3.1 for the year 2006 can be examined in conjunction with petitions filed with the

National Vaccine Compensation Injury Program that same year when 325 claims were filed (not adjudicated) (Health Resources and Services Administration, 2017). This number must be put into context of the sheer volume of vaccines administered:

> According to the CDC, from 2006 to 2015 over 2.8 billion doses of covered vaccines were distributed in the U.S. For petitions filed in this time period, 4,374 petitions were adjudicated by the Court, and of those 2,847 were compensated. This means for every 1 million doses of vaccine that were distributed, 1 individual was compensated. (Health Resources and Services Administration, 2017: 1)

Compensated claims do not equate to the determination that the vaccine caused injury. Instead, most compensation is the result of negotiated agreements where it is unclear if the vaccine was the cause of the injury (Health Resources and Services Administration, 2017).

Preventing the reemergence of diseases is an important public health concern. With hospitals now seeing cases of VPDs in patients, antivaccinators are feeling pressure to comply with vaccination schedules for their children to ensure (as much as possible) herd immunity. Although the connection between vaccines and autism or other neurological impacts has been widely debunked, persistent voices, such as Robert F. Kennedy Jr.'s, who continue to focus on the dangers of vaccines, instill doubts in vaccine-hesitant individuals. Here is what Robert F. Kennedy Jr. wrote in a recent "manifesto" about vaccines:

> Vaccines are big business. Pharma is a trillion-dollar industry with vaccines accounting for $25 billion in annual sales. CDC's decision to add a vaccine to the schedule can guarantee its manufacturer millions of customers and billions in revenue with minimal advertising or marketing costs and complete immunity from lawsuits. High stakes and the seamless marriage between Big Pharma and government agencies have spawned an opaque and crooked regulatory system. (2017)

While Kennedy claims to be provaccine, he bases arguments against the vaccine not on science but on the potentially corrupt relationship between pharmaceutical companies and government. In response, an article in the *Atlantic* counters Kennedy's claim, stating that the Big-Pharma/big-money argument is "historically unfounded" (Lam, 2015). Further, "not only do pediatricians and doctors often lose money on vaccine administration, it wasn't too long ago that the vaccine industry was struggling with slim profit margins and shortages" (Lam, 2015). Irrespective of the mounting evidence showing that vaccines do not cause autism and that only in very rare cases are there vaccine injuries, people's desire to protect their own child even at the expense of VPDs reemerging in society is a strong motivator for some to disregard science. Instead, these individuals favor the minority of voices insisting that there is vaccine harm where in fact there is none or suggest a conspiracy where one does not exist.

VACCINE SKEPTICS

To be sure, those refusing all vaccinations (antivaccinators) are in a small minority. But that minority can have a huge impact on public health. So who are the vaccine skeptics or antivaccinators? A 2015 report by the Pew Research Center concerning views about vaccinations found that age, ideology, and political party affiliation heavily influence opinions on vaccinations for children (Pew Research Center, 2015a). Specifically, older people and Democrats (or those with a Democratic leaning) are more supportive of requiring childhood vaccinations. However, the Pew study did not inquire about vaccine practices. In this regard, other studies have found that white women over the age of 30 (Smith et al., 2004; Gust et al., 2008; Salmon et al., 2015), particularly those with higher education levels and income (Smith et al., 2004; Salmon et al., 2015), were more likely to abstain from vaccinating their children. Research also shows regional differences with one study finding a greater proportion of antivaccinators in the western part of the United States (Gust et al., 2008) and another finding cluster groups more widespread throughout the United States, in California, Colorado, Illinois, Michigan, New York, Oklahoma, Pennsylvania, Texas, Utah, and Washington State (Smith et al., 2004). Whereas people once trusted their doctors for vaccine recommendations, many people now frequently use the Internet to obtain online health information, where conflicting "evidence" about vaccines is abundant.

In a 2012 study, Abbey M. Jones et al. found that parents seeking information about vaccinations via the Internet (where antivaccine messages are prevalent) were less likely to believe in vaccine science, were more likely to obtain nonmedical exemptions, and were more likely to trust the advice of nontraditional health care providers (e.g., alternative or holistic) as a reliable source. In another study, Mergler et al. (2013) suggested a relationship between antivaccination beliefs and choice of medical provider. Parents who express concern over weakening their child's immune system due to vaccinations "had 4.6 higher odds of having a provider who shared this belief compared to parents whose providers did not agree or strongly agree with this belief" (Mergler et al., 2013: 4593.)

Jenny McCarthy, Vaccine Safety Advocate

I do believe sadly it's going to take some diseases coming back to realize that we need to change and develop vaccines that are safe.

—Jenny McCarthy

Autism is a neurological disorder that is often diagnosed around two years of age. Autism impacts a child's social interaction, including verbal and nonverbal communication. Particularly heartbreaking is that often children seem typical and exhibit regular milestone behavior before autistic behaviors begin. Autism is often diagnosed around the same time as the first MMR sequence is delivered (between 12 and 15 months) in conjunction with other routine vaccinations. Parents searching for a reason for the autism diagnosis suggest that vaccinations are causally associated with autism. One champion of this view is celebrity Jenny McCarthy. McCarthy's own son was diagnosed as autistic, sending her on a personal journey to identify the cause of her son's diagnosis. McCarthy is now the author of several books on autism, a vocal champion for autistic causes, and the president of Generation Rescue, an autism recovery organization. While McCarthy offers many people of autistic children comfort and hope, the messaging that vaccines cause autism without evidence to support such a claim has undoubtedly instilled reservation among parents about vaccinating their children. In reality, certain risk factors are key to understanding an autistic diagnosis such as: genetic and chromosomal disorders, higher parent age at birth, premature birth, and if an identical twin has autism (CDC, 2017c). Certainly, it is every parent's wish to protect his or her child from harm. However, in the case of vaccines, there is abundant proof that without vaccination children are susceptible to incapacitating and deadly diseases, and no proof that vaccines cause neurological harm.

It appears that due to selective bias, parents who abstain from vaccinating their kids tend to reinforce their choice by putting greater trust in health care providers who share their view, selecting information via the Internet that "validates" antivaccination beliefs, and believing in the celebrity champions (like Jenny McCarthy) who are outspoken antivaccination advocates prevalent on the Internet (Gowda and Dempsey, 2013).

Finally, while there are pockets of unvaccinated children in several states, coincidentally often in states that allow for philosophical exemptions for vaccines (Smith et al., 2004), it is primarily the western region (Gust et al., 2008; Smith et al., 2004) that reports the highest numbers of unvaccinated children. Further, cluster groups of unvaccinated children are related to the role of social networks. Specifically, social networks—including spouses or partners, friends, teachers, midwives, and others—play a strong role in antivaccination decisions (Brunson, 2013), helping to explain the geographic clusters of unvaccinated children.

IDEOLOGY, POSITIVISM, AND POSTMATERIALISM

There exists conflicting information about antivaccinators and vaccine skeptics. Is vaccine skepticism the domain of antiscience liberals? Or is this an issue that transcends ideology and rests more on beliefs in science and postmaterialist values? Certainly, understanding who opposes (or questions) vaccinations is vital in addressing matters affecting public health and safety.

Ideology: The ideological underpinnings behind antivaccination choices perpetuate vaccine hesitancy or rejection through the denial of scientific consensus, which resonates within a certain segment of the U.S. population. Prior studies have shown that various sociodemographics are related to antivaccination choices (e.g., liberal, older white mothers, higher income and education, and geographic pockets). But vaccine skepticism may have support among (and even be driven by) conservatives as well. Recently, Donald Trump appointed Dr. Tom Price as secretary of health and human services. Price belongs to the Association of American Physicians and Surgeons, a conservative group with arguably opposite views from mainstream federal health policy (Goldstein, 2017) and one that "opposes mandatory vaccination as 'equivalent to human experimentation,' a stance contrary to requirements in every state" (Goldstein, 2017) and against mainstream medical consensus on vaccines. So when asked whether vaccines like MMR should be required, Dr. Price responded, "I believe it's a perfectly appropriate role for the government, this happens by and large at the state-government level, because they're the ones who have the public-health responsibility . . . to determine whether or not immunizations are required for a community population" (Levitz, 2017).

While it is true that states determine the laws pertaining to requirements for vaccines, diseases do not abide by state borders. This statement, in conjunction with Trump's comments on vaccines and autism, and Price's affiliation with a group that is decidedly antivaccination, has "energized the anti-vaccine movement" with the movement "becoming more popular, raising doubts about basic childhood health care among politically and geographically diverse groups" (Sun, 2017).

Just how much does political ideology influence vaccination decisions? On certain policy issues, like climate change, the liberal-conservative divide is obvious. Conversely, vaccination attitudes seem to capture people on the far edges of the ideological spectrum. When liberals like Robert F. Kennedy Jr. find common ground with conservatives like Donald Trump and Tom Price, vaccine skepticism appears more ideologically distributed than literature suggests.

Several recent studies have found that antivaccination attitudes may be more prevalent in conservative circles. A study conducted by Lupton and Hare (2015) determined that "the probability of believing in a link between vaccines and autism is much higher among conservatives than liberals—regardless of whether people identified as Democrats, Republicans, or independents." A 2017 study by the Pew Research Center found that conservatives are slightly more likely to leave vaccine choices up to parents, even at the expense of community health (Funk, Kennedy, and Hefferon, 2017). And vocal opposition from conservatives like Michele Bachmann surfaced regarding the HPV vaccine (in part due to the HPV vaccine's aim of immunizing teenagers from an STD).

Considering that the pharmaceutical industry is the fifteenth-largest donor to the 2016 presidential election and spent millions of dollars lobbying for favorable health care policies (Kounang, 2016), it suggests potential evidence of big business's influence over government decisions. It is potentially this corporate-government connection that explains why only 13 percent of U.S. adults trust health information about the MMR vaccine from pharmaceutical companies (Funk, Kennedy, and Hefferon, 2017). As liberals are even less likely than conservatives to support research from for-profit organizations, especially those that challenge science to maintain or increase profits by undermining federal regulatory agencies (McCright and Dunlap, 2003), the ideological divide over vaccines becomes murky.

The development and implementation of vaccines are the shared domain of pharmaceutical companies and the government. It is the merger of these entities that potentially aligns opinions from both conservatives and liberals on the rejection of vaccines. A study by Lewandowsky, Gignac, and Oberauer found that "opposition to vaccination involved a balance between two opposing forces, namely a negative association with free-market endorsement and a compensatory positive association with conservatism" (2013: e75637). Therefore conservatives who oppose government intervention (and mandates), and liberals who are skeptical of pharmaceutical companies may find common ground in vaccine opposition.

Positivism: Studies concerning antivaccination activists focus on themes of trust: trust that the vaccine is safe, trust that it is effective, trust in the companies manufacturing the vaccines and government agencies ensuring safety, and trust in scientists and doctors. At the core of these issues of trust is a belief, or lack thereof, in science to provide truths about the world around us. A positivistic scientific position holds that the scientific method and empirical data are critical in establishing scientific consensus by providing observable, replicable findings. In this view, science is the most reliable way to understand the world and through strict methodological research can identify facts or truths (Steel et al., 2004). Using a positivist lens, it is through science that we can gain deeper understanding about medicine, like vaccines, which provides benefits to society by improving overall health and well-being.

While many turn toward science to reveal deeper knowledge about the world, research finds that in many areas there are significant divides between issues with relative scientific consensus and public attitudes. In a 2015 Pew study, researchers found that there was an 18 percent gap between the public and scientists on the issue of requiring the MMR vaccine, (68% public to 86% scientists) (Pew Research Center, 2015a). The lack of public knowledge, and potentially trust, in science concerns scientists, with 84 percent saying that the "public doesn't know much about science" (Pew Research Center, 2015a). Regarding vaccine skeptics, this is true. A Pew report found that "people with low knowledge about science are also less likely to see high preventative health benefits from vaccines" (55% compared with 91% of those with high science knowledge) (Funk, Kennedy, and Hefferon, 2017). Conversely, "people with high science knowledge are especially positive in their views of medical science and research on childhood vaccines" (Funk, Kennedy, and Hefferon, 2017).

While science is not without error, lack of trust in (or denial of) scientific findings, such as the safety of vaccines, suggests that it is not simply paucity of knowledge but perhaps instead a confirmation bias toward alternative "evidence" finding vaccines to be unsafe. People who do not ascribe to a positivistic model of science may therefore be skeptical of vaccine safety, efficacy, and necessity. Basically, "if a scientific consensus cannot be accepted as the result of researchers converging independently on the same evidence-based view, then the belief in a scientific conspiracy can provide an alternative explanation for the consensus" (Lewandowsky, Gignac, and Oberauer, 2013: e75637).

Postmaterialism: The manifestation of postmaterialist values began in earnest in the 1960s with the environmental movement, the peace movement, and more broadly awareness and attention to issues impacting quality of life (Steel et al., 1992). Conis (2015) suggests that the current antivaccination movement is an outgrowth of many of the social movements of the 1960s and 70s, particularly the environmental movement, the feminist movement, and the child protection movement. Social movements like the environmental and feminist movements focused a great deal of attention either directly or indirectly on human health and choice. As information about the negative impacts of DDT emerged with Rachel Carson's *Silent Spring* (1962), people began to consider toxins and their possible adverse impact on human health, particularly in vulnerable populations like children.

In a democratic postmaterial nation, like the United States, the political system allows more points of access for policy influence. The antivaccination movement gained traction by utilizing the access points created via the Internet. The movement found it easy to distribute information through the Internet and created a base of support for vaccine refusal and an avenue to pursue vaccine policy reforms. In this way, technology paired well with the postmaterial "ethic that consists of such enlightened values as freedom of access, sharing to the benefit of others, [and] using technology to improve the world" (Lehdonvirta, 2010: 884).

Hence, in the larger issue of vaccines, technology both served to protect the general population from VPDs and allowed greater distribution of information contradicting U.S. vaccine policies. With greater access to information via the Internet, skepticism about the efficacy and safety of vaccines is abundant. Approximately 74 percent of U.S. adults use the Internet, with roughly 80 percent of the users having used it for health-related information (Fox, 2011). Of those seeking online health information, 52 percent believe that the information they receive online is credible (Rainie, 2000). Considering that a Google search of "vaccination" results in 71 percent antivaccination sites (Kata, 2010), the Internet is an effective tool for distributing information for the antivaccination movement. This may be one reason that research has found an increase in concern pertaining to vaccines. In 2000, 19 percent of parents in a national study indicated they were concerned about vaccines, compared to 50 percent of parents indicating concern in 2009 (Gowda and Dempsey, 2013).

The outputs of postmaterialism from social movements and technology have created a more discerning public and, one could argue, a public more skeptical of big business and government. With the imminent threats of VPDs lessened, antivaccination groups have centralized their messages around four predominant themes: safety and efficacy; alternative, holistic medicine to prevent disease; conspiracy theories; and civil liberties arguments (Kata, 2010; Blume, 2006). Thus antivaccination groups "make postmodern arguments that reject biomedical and scientific 'facts' in favor of their own interpretations" (Kata, 2010: 1709).

ANALYSES

In order to examine public attitudes about vaccinations, the West Coast survey focused on six vaccination statements using a Likert scale where 1 = strongly disagree and 5 = strongly agree. Several of the statements included in the survey were based on a poll conducted by the Pew Research Center in 2015 (Anderson, 2015) (see Table 3.2). For the first statement, "some vaccines can cause autism in children," we find that 9.6 percent strongly agree and 12.2 percent agree. This is 21.8 percent total taking an antivaccination position, which is somewhat higher than what the 2015 Pew Research Center found. Half of the respondents disagreed or strongly disagreed with the statement and about a little over one-fourth of respondents were neutral.

For the statement "there are serious side effects from vaccinations," we find 17.0 percent strongly agreeing and 18.5 percent agreeing, which is over one-third of total survey respondents. A little over 40 percent disagreed or strongly disagreed with the statement and 24.2 percent were neutral. Interestingly, only 6.8 percent of respondents agreed and strongly agreed with "if other people vaccinate their children, then I do not need to vaccinate my family."

For the final three statements included in the survey, we find skewed results in favor of vaccinations, just as the 2015 Pew Research Center survey also found. When asked to respond to the statement "I feel there is little risk from vaccinations," 17.2 percent of respondents disagreed and strongly disagreed, compared with 66 percent agreeing and strongly agreeing. For the fifth statement, "It is necessary for everyone to get vaccinated to protect community health," 16.3 percent disagreed or strongly disagreed and 68.7 percent agreed and strongly agreed. For the final statement, "I feel vaccines are safe," we find that 12.3 percent of respondents disagreed and strongly disagreed while 66.3 percent agreed and strongly agreed.

For the six statements, antivaccine attitudes ranged from 6.8 percent to 35.5 percent, and provaccine attitudes ranged from 40.3 percent to 88.8 percent. As noted above, the percentage of antivaccination responses was somewhat higher than the 2015 national Pew Research Center survey. That survey found "roughly eight-in-ten U.S. adults (83 percent) say vaccines . . . are safe for healthy children" (Anderson, 2015). However, it may also be that the case study states have higher

Table 3.2 Public Attitudes toward Vaccinations

Question: How likely are you to agree or disagree with the following statements about immunizations (vaccinations)? [1 = Strongly disagree to 5 = Strongly agree]

Variable Name		Strongly Disagree *Percent*	Disagree *Percent*	Neutral *Percent*	Agree *Percent*	Strongly Agree *Percent*
Vaccine 1	Some vaccines can cause autism in children [N = 1,479]	25.3	25.1	27.9	12.2	9.6
Vaccine 2	There are serious side effects from vaccinations [N = 1,478]	19.5	20.8	24.2	18.5	17.0
Vaccine 3	If other people vaccinate their children, I do not need to vaccinate my family [N = 1,484]	48.0	40.0	5.2	3.7	3.1
Vaccine 4	I feel there is little risk from vaccinations [N = 1,484]	8.3	8.9	16.8	35.3	30.7
Vaccine 5	It is necessary for everyone to get vaccinated to protect community health [N =1,479]	7.2	9.1	14.7	32.5	36.5
Vaccine 6	I feel vaccines are safe [N = 1,483]	6.3	6.0	21.4	34.7	31.6

percentages of people who are antivaccine than other states. For example, one recent study found large clusters of politically liberal communities in Northern California with high rates of immunization refusals (Ingraham, 2015). Similarly, other studies have identified antivaccination communities in Washington State (e.g., Vashon Island in Puget Sound) and Oregon (certain neighborhoods in Portland and Ashland) where large numbers of "counterculture" residents can be found (Peeples, 2015).

The data presented in Table 3.3 present correlations between the six immunization questions to examine the consistency of responses.

Table 3.3 Correlation Coefficients for Vaccination Orientations

Question: How likely are you to agree or disagree with the following statements about immunizations (vaccinations)?

		Vaccine 1	Vaccine 2	Vaccine 3	Vaccine 4	Vaccine 5	Vaccine 6
			Tau b	*Tau b*	*Tau b*	*Tau b*	*Tau b*
Vaccine 1	Some vaccines can cause autism in children [N = 1,474 to 1,479]		.426**	.297**	−.441**	−.395**	−.492**
Vaccine 2	There are serious side effects from vaccinations [N = 1,473 to 1,478]			.108**	−.424**	−.336**	−.423**
Vaccine 3	If other people vaccinate their children, I do not need to vaccinate my family [N = 1,479 to 1,484]				−.232**	−.384**	−.342**
Vaccine 4	I feel there is little risk from vaccinations [N = 1,479 to 1,484]					.369**	.615**
Vaccine 5	It is necessary for everyone to get vaccinated to protect community health [N = 1,479]						.544**

** $p \leq .01$ (one-tailed)

The results indicate that all of the correlations are statistically significant at .01, with the strength of the relationships ranging from .108 to .615. The direction of the relationships shows consistency in orientations with the first three antivaccination statements being negatively correlated with the final three statements. For example, those respondents that agreed with "some vaccines can cause autism in children" were significantly likely to disagree with the statements "I feel there is little risk from vaccinations," "it is necessary for everyone to get vaccinated to protect community health," and "I feel vaccines are safe."

The next set of tables examines bivariate relationships between ideology, postmaterialist values, and beliefs about positivism, similar to what was presented in Chapters 1 and 2. Table 3.4 presents bivariate results for the impact of political ideology on vaccination orientations. While political ideology has a statistically significant impact for all of the six immunization statements, the pattern is not always clearly liberal versus conservative, which is consistent with past research. For example, 58.5 percent of liberals disagree that vaccinations cause autism in children compared to 43.7 percent of conservatives; 23 percent of liberals agreed with the statement compared to 18.1 percent of conservatives. However, for other statements, there were more noticeable differences. For example, 40.7 percent of liberals agreed that "there are serious side effects from vaccinations" compared to 25.1 percent of conservatives. Similarly, conservatives were more likely than liberals to believe that "there is little risk from vaccinations" and that "vaccines are safe."

In terms of the impact of postmaterialist values on vaccination orientations, the data displayed in Table 3.5 are similar to the impact of political ideology with some mixed results. While all of the Chi-square statistics are significant, postmaterialists are somewhat similar in their level of agreement with materialists for the statements "vaccinations cause autism in children" and "if other people vaccinate their children, I do not need to vaccinate my family." However, the expected pattern of postmaterialists having more antivaccination orientations when compared to other value types holds for: "there are serious side effects from vaccinations," "there is little risk from vaccinations," "it is necessary for everyone to get vaccinated to protect community health," and "vaccines are safe."

Finally, when examining the impact of belief in positivism, we find the expected pattern of less positivistic people having greater levels of antivaccination orientations than those with higher levels of belief in positivism (see Table 3.6). Over 31 percent of respondents with low levels of belief in positivism agree that vaccinations cause autism compared to 13.2 percent of those with high levels of belief. Similarly, 55.9 percent of those with low levels of belief agree that there are serious side effects from vaccinations when compared to 17.3 percent of high-level believers. In addition, those with low levels of belief in positivism were significantly

Table 3.4 Political Ideology and Vaccination Orientations

a. Some vaccinations can cause autism in children [Chi-square = 61.898, $p = .000$]

	Liberal *Percent*	Moderate *Percent*	Conservative *Percent*
Disagree	58.5	42.9	43.7
Neutral	18.5	33.6	38.2
Agree	23.0	23.5	18.1
N =	687	387	398

b. There are serious side effects from vaccinations [Chi-square = 31.811, $p = .000$]

	Liberal *Percent*	Moderate *Percent*	Conservative *Percent*
Disagree	38.7	36.1	47.2
Neutral	20.6	27.6	27.6
Agree	40.7	36.3	25.1
N =	685	388	398

c. If other people vaccinate their children, I do not need to vaccinate my family [Chi-square = 32.563, $p = .000$]

	Liberal *Percent*	Moderate *Percent*	Conservative *Percent*
Disagree	83.5	94.1	89.7
Neutral	6.7	4.1	3.8
Agree	9.9	1.8	6.5
N =	689	390	398

d. I feel there is little risk from vaccinations [Chi-square = 33.247, $p = .000$]

	Liberal *Percent*	Moderate *Percent*	Conservative *Percent*
Disagree	21.2	13.3	14.1
Neutral	13.6	24.4	14.6
Agree	65.2	62.3	71.4
N =	689	390	398

e. It is necessary for everyone to get vaccinated to protect community health [Chi-square = 73.703, $p = .000$]

	Liberal *Percent*	Moderate *Percent*	Conservative *Percent*
Disagree	19.9	9.3	16.4
Neutral	10.0	26.8	10.6
Agree	70.1	63.9	73.0
N =	688	388	397

Table 3.4 (Continued)

f. I feel vaccines are safe [Chi-square = 41.562, $p = .000$]

	Liberal *Percent*	Moderate *Percent*	Conservative *Percent*
Disagree	12.6	9.0	14.3
Neutral	25.7	24.6	10.6
Agree	61.7	66.4	75.1
N =	689	390	398

Table 3.5 Postmaterialist Values and Vaccination Orientations

a. Some vaccinations can cause autism in children [Chi-square = 75.717, $p = .000$]

	Postmaterialist ***Percent***	**Mixed** ***Percent***	**Materialist** ***Percent***
Disagree	43.7	53.0	39.4
Neutral	38.2	32.2	39.4
Agree	18.1	14.8	21.2
N =	553	860	66

b. There are serious side effects from vaccinations [Chi-square = 85.477, $p = .000$]

	Postmaterialist *Percent*	**Mixed** ***Percent***	**Materialist** ***Percent***
Disagree	31.6	46.6	31.8
Neutral	18.7	27.5	27.3
Agree	49.7	25.9	40.9
N =	551	861	66

c. If other people vaccinate their children, I do not need to vaccinate my family [Chi-square = 68.684, $p = .000$]

	Postmaterialist *Percent*	**Mixed** ***Percent***	**Materialist** ***Percent***
Disagree	79.9	93.4	84.4
Neutral	8.9	3.2	0.0
Agree	11.2	13.4	15.2
N =	553	865	66

(Continued)

Table 3.5 (Continued)

d. I feel there is little risk from vaccinations [Chi-square = 72.490, $p = .000$]

	Postmaterialist *Percent*	Mixed *Percent*	Materialist *Percent*
Disagree	26.9	12.3	0.0
Neutral	17.9	15.7	22.7
Agree	55.2	72.0	77.3
N =	553	865	66

e. It is necessary for everyone to get vaccinated to protect community health [Chi-square = 30.662, $p = .000$]

	Postmaterialist *Percent*	Mixed *Percent*	Materialist *Percent*
Disagree	22.2	12.6	15.2
Neutral	12.1	17.1	6.1
Agree	65.6	70.3	78.8
N =	553	388	66

f. I feel vaccines are safe [Chi-square = 78.621, $p = .000$]

	Postmaterialist *Percent*	Mixed *Percent*	Materialist *Percent*
Disagree	19.0	7.3	21.2
Neutral	27.3	18.2	15.2
Agree	53.7	74.5	63.6
N =	553	864	66

Table 3.6 Positivism Beliefs and Vaccination Orientations

a. Some vaccinations can cause autism in children [Chi-square = 69.197, $p = .000$]

	Low *Percent*	Medium *Percent*	High *Percent*
Disagree	40.9	48.3	63.4
Neutral	27.8	31.7	23.5
Agree	31.4	19.9	13.2
N =	526	482	456

Table 3.6 (Continued)

b. There are serious side effects from vaccinations [Chi-square = 199.558, $p = .000$]

	Low *Percent*	Medium *Percent*	High *Percent*
Disagree	26.4	35.8	59.4
Neutral	17.7	32.9	23.2
Agree	55.9	31.3	17.3
N =	526	480	456

c. If other people vaccinate their children, I do not need to vaccinate my family [Chi-square = 38.729, $p = .000$]

	Low *Percent*	Medium *Percent*	High *Percent*
Disagree	83.3	90.7	90.1
Neutral	10.0	2.3	2.9
Agree	6.6	7.0	7.0
N =	528	484	456

d. I feel there is little risk from vaccinations [Chi-square = 111.561, $p = .000$]

	Low *Percent*	Medium *Percent*	High *Percent*
Disagree	29.5	7.9	13.4
Neutral	16.3	23.1	11.4
Agree	54.2	69.0	75.2
N =	528	484	456

e. It is necessary for everyone to get Vaccinated to protect community health [Chi-square = 93.850, $p = .000$]

	Low *Percent*	Medium *Percent*	High *Percent*
Disagree	27.5	11.2	9.4
Neutral	16.0	17.8	9.6
Agree	56.5	71.0	80.9
N =	524	483	456

f. I feel vaccines are safe [Chi-square = 101.247, $p = .000$]

	Low *Percent*	Medium *Percent*	High *Percent*
Disagree	20.1	8.1	8.1
Neutral	27.5	25.1	11.4
Agree	52.5	66.9	80.5
N =	528	483	456

more likely than those with high levels of belief to have antivaccination orientations for "risk from vaccinations," vaccinations for community health, and disagreeing that vaccines are safe.

The final analyses conducted with the survey data concern whether several sociodemographic characteristics, political ideology, postmaterialist values, and attitudes toward science are associated with the antivaccination archetype as discussed above. We address this with a multivariate analysis of the six statements concerning the risk and safety of vaccinations presented in Table 3.2. As with the multivariate analyses conducted in Chapter 2 for GMOs, the sociodemographic variables examined include age in years, gender, and formal educational attainment.

Table 3.7 presents results for six ordinary least squares regression models examining the validity of our antivaccination archetype. F-test results for all six models are statistically significant, indicating a relatively good fit overall. Adjusted R^2 results range from a low of .035 for the Vaccine 3 model to a high of .146 for the Vaccine 4 model. Of course the distribution of results for the "if other people vaccinate their children, I do not need to vaccinate my family" statement were heavily skewed toward disagreement, leaving little variation to explain them to begin with (see Table 3.2). Therefore some of the findings suggest there is much more to explore to account for variation in levels of antivaccine sentiment (e.g., Vaccine 3 and 5 models), while several models show more robust results (i.e., Vaccine 2, 4, and 6).

All three sociodemographic variables were found to be predictors of certain antivaccine positions. Age was statistically significant in all six models with younger respondents more likely supporting antivaccine positions when compared to older respondents. Gender was found to be significant in only one model, with females more likely to take an antivaccination position when compared to males concerning the safety of vaccines (Vaccine 6 model). Similar to the impact of age, formal educational attainment had a statistically significant impact in all six models. More specifically, support for antivaccine positions decreases among those with higher levels of education. This finding may suggest that the highly educated are more likely to access information sources associated with mainstream, peer-reviewed science.[6]

Turning now to our value orientations, we find that the additive index assessing positivist orientations toward science (positivism) has a statistically significant impact in five of the six models. The positivism indicator did not have a statistically significant impact for the Vaccine 3 model. However, for the remaining models, we find that those with lower levels of support for positivism were significantly more likely to have

Table 3.7 Regression Estimates for Antivaccination Orientations

Variable	Vaccine 1 *Coefficient* *(Std. Error)*	Vaccine 2 *Coefficient* *(Std. Error)*	Vaccine 3 *Coefficient* *(Std. Error)*	Vaccine 4 *Coefficient* *(Std. Error)*	Vaccine 5 *Coefficient* *(Std. Error)*	Vaccine 6 *Coefficient* *(Std. Error)*
Age	–.011***	–.017***	–.006***	.014***	.008***	.010***
	(.002)	(.002)	(.002)	(.002)	(.002)	(.002)
Gender	.122	.035	.051	.014	–.090	–.116*
	(.064)	(.068)	(.051)	(.062)	(.064)	(.058)
Education	–.162***	–.076***	–.053***	.095***	.055**	.110***
	(.021)	(.022)	(.016)	(.020)	(.021)	(.019)
Ideology	–.076***	–.010	.001	.000	–.029	.003
	(.017)	(.018)	(.013)	(.016)	(.017)	(.015)
Postmaterialist	.141*	.128*	.167***	–.263***	–.073*	–.207***
	(.058)	(.061)	(.046)	(.056)	(.058)	(.052)
Positivism	–.042***	–.080***	–.008	.056***	.056***	.052***
	(.006)	(.007)	(.005)	(.006)	(.006)	(.006)
F-test =	36.954***	53.327***	9.074***	42.454***	25.406***	40.734***
Adjusted R^2 =	.130	.179	.035	.146	.092	.141
N =	1,446	1,444	1,450	1,444	1,440	1,444

*$p \leq .05$; **$p \leq .01$; ***$p \leq .001$

antivaccine positions compared to those with higher levels, who had more support for vaccination orientations. We argued above that provaccination advocates would be more likely to use scientific studies and evidence of VPDs when compared to antivaccination advocates. While we did not assess the types of information sources people use for this issue, those with high levels of belief in vaccinations would at least be more open to the scientific literature on this topic.

The indicator of political ideology was statistically significant for only one model—the view that vaccines cause autism in children (Vaccine 1). As suggested in our discussion above, many antivaccine advocates and communities are associated with the left and/or countercultures; however, there is evidence suggesting that conservatives may also object to vaccinations, confirming why the measure of political ideology used in this study did not predict immunization attitudes well while controlling for sociodemographic and other values orientations.

The final variable included in each is the measure of postmaterialist values. As we discussed earlier, Inglehart labeled this set of values "postmaterialism" because the values experienced are "beyond" the materialistic concerns of the lower levels of the need hierarchy that dominated the preceding era. As the discussion above predicted, the postmaterialist dummy variable was statistically significant in all six models. Postmaterialists when compared to other values types (i.e., "mixed" and "materialist" values) are significantly more likely to take an antivaccination position for Vaccine 1 (vaccines cause autism), Vaccine 2 (vaccines have serious side effects), Vaccine 3 (other people vaccinate, I do not need to vaccinate my family), Vaccine 4 (little risk from vaccinations), Vaccine 5 (necessary for everyone to get vaccinated), and Vaccine 6 (vaccines are safe).

The public survey results examined here found support for some of the attributes the literature suggested. We found that the demographic variable of gender had little impact for most vaccination attitudes with the exception of the statement regarding safety. However, both age and formal educational attainment were robust predictors of vaccination attitudes with the younger and lower education levels being significantly more likely to have antivaccination attitudes.

Finally, in terms of the major focus of this book, values do have an impact on vaccination attitudes. Postmaterialists are significantly more likely to take antivaccination positions when compared to other value types. In addition, those respondents who have lower levels of belief in a positivistic science were also significantly more likely to have antivaccination attitudes when compared to those with higher levels of belief that science can be positivistic.

CONCLUSION

In terms of overall vaccine knowledge, there are some valuable, if not concerning, trends and inconsistencies. Results suggest a strong knowledge gap pertaining to the safety and efficacy of vaccinations. The resurgence of vaccine skepticism in popular media, the Internet, and our current political climate has most likely instilled, at minimum, seeds of doubt about vaccine safety and necessity. The unfortunate popularity of the overhyped, and erroneous, Wakefield study that implied a link between vaccines (MMR, specifically) and autism is repeatedly cited in many news stories and antivaccination communities although the study has been discredited. The Wakefield study, anecdotal cautionary tales (like those perpetuated by President Trump), along with other "evidence" of vaccine injuries, while not scientifically rigorous, have potentially cast doubt on vaccines that ultimately protect human health. As VPDs continue to sporadically crop up in areas around the United States, it is becoming increasingly important to counteract the antivaccine narrative to maintain safeguards against debilitating and deadly diseases.

This chapter focused on key individual variables that are potentially predictive of people who chose not to vaccinate or intentionally undervaccinate. Our findings suggest that antivaccination attitudes and beliefs are complex. Popular media often describe antivaccinators as well-educated, liberal, white women. However, this description of antivaccinators perpetuates inaccuracies that might make it more challenging to identify those who undervaccinate their children or chose to abstain from vaccinations.

One of the most consistent findings in academic literature on antivaccinators is gender. Although some mothers are now more educated and working outside of the home, division of labor related to child care remains primarily with the mother. A 2014 study found that roughly three-quarters of mothers reported that they handled all the primary medical care issues for children (finding doctors, taking children to doctor's appointments, etc.) (Ranji and Salganicoff, 2014). In another study, in homes with two working parents, the majority responded (55%) that women do more of management of children's schedule and taking care of kids when they are sick (Pew Research Center, 2015c). Given that women are often the primary caregiver for children, it would follow that they are the ones most likely to make vaccine decisions and seek information to help form decisions on vaccines. Although women may still indeed make decisions pertaining to children's medical care, our study found that gender had little influence over vaccination attitudes.

Gender plays only a small role on attitudes toward vaccine safety, where women are more likely to express hesitancy.

Many antivaccine narratives are predominantly featured on Internet vaccine searches, and findings from a 2009 study found that "many parents are using, and place at least some trust in, a variety of Web sites for vaccine safety information, including those from groups that oppose vaccines" (Freed et al., 2010). The study further found that women are more likely to trust nonprofessional sources of vaccine information, and that 26 percent place trust in vaccine information from celebrities and 73 percent trust parents who share stories about their child's vaccine injuries (Freed et al., 2010). While this same study found that 76 percent of parents trust their child's doctor for vaccine information (Freed et al., 2010), parents also seek medical providers who tend to reinforce their beliefs regarding vaccinations (Mergler et al., 2013; Jones et al., 2012). Collectively, this research reinforces our findings that people with weaker positivistic views demonstrate more antivaccination attitudes.

Further, we found antivaccination attitudes consistent with postmaterialist values. Postmaterialist values are more "liberal" and have been associated with the liberal policy causes of the 1960s, '70s, and '80s—such as civil rights, antiwar and antinuclear activism, gender equality, protection of the environment, and the advocacy of more "open" lifestyles. We suggested above that these values might be associated with antivaccination positions because of postmaterialism's liberal tendencies in terms of civil liberties, and support the ability of individuals and families to make their own decisions about vaccinations and other lifestyle choices. Conversely, while liberalism is often associated with postmaterialist values, ideology was not overall a significant predictor of antivaccination attitudes, except regarding concern that vaccines cause autism where liberals were more likely to express a perceived correlation between vaccines and autism.

Overall, antivaccination attitudes ranged from 7.9 percent to 36.7 percent. Although provaccination attitudes were much higher, 40.5 percent to 86.7 percent, if people indeed acted on their antivaccination attitudes, vaccine prevention for communities, states, and even the nation could diminish. As mentioned earlier, our sampling of the West Coast may represent slightly higher antivaccination attitudes (Gust et al., 2008) but nonetheless is significant if, at minimum, 90 percent of the population needs to be immunized to provide protection against VPDs.

We find that people are more likely to support a provaccination position if they have higher education, a stronger positivistic position, and are more conservative ideologically. People more likely to support an antivaccination position are less positivistic, younger, identify as a postmaterialist, and have lower education levels. Based on our findings, antivaccination attitudes are not easily confined to a certain "group" of people but reflect the complexity of identifying antivaccinators or people with certain vaccine hesitancies.

This study has several drawbacks. First, respondents were not asked about vaccination behaviors, only attitudes. Therefore there is nothing to directly suggest whether people with negative vaccination attitudes acted on this by not vaccinating or intentionally undervaccinating their children. Second, questions pertaining to race/ethnicity or income were also not included in the survey. While both of these have been indicators of antivaccination attitudes in prior research, they are not addressed in this study. Finally, respondents were not asked where they obtain information on vaccines. Because the antivaccination narrative is predominant on television, the Internet, and other media outlets (newspapers, magazines), it would be interesting to determine what news sources antivaccinators trust most, or if information is obtained directly from a medical professional (traditional or alternative).

Successful vaccination programs depend greatly on high levels of vaccination; therefore it is important to dispel myths about those who chose not to vaccinate or intentionally undervaccinate their families. This research focused on whether variables like sociodemographics, political ideology, postmaterialism, and positivism affect attitudes toward vaccinations. We found that, somewhat contrary to popular belief, antivaccination attitudes are aligned with younger people, those with lower levels of education, people who are less positivistic, and those with stronger postmaterial views.

Because public health is contingent on very high levels of vaccinations, the broader public policy question is how to gain compliance with people who are vaccine hesitant or antivaccination. Ultimately, it may be difficult, if not impossible, to convince those with ideological opposition to vaccines (and as prior research suggests, may have the opposite effect of further solidifying antivaccination attitudes). Since many people trust their medical providers with vaccine choices, the key may be to work with both traditional and nontraditional (holistic) medical providers to ensure that patients are receiving accurate vaccine information and that myths pertaining to vaccine injuries are dispelled.

The other policy option is to remove the philosophical and ideological exceptions for unvaccinated children in public schools. Currently, several states allow for parents to enroll their unvaccinated children in public schools if they file an exception. Although this would be an unpopular, if not impossible, policy option, it would potentially increase vaccination rates among children attending public schools.

Vaccines have prevented over 6 million deaths worldwide, and combined with clean water, has been one of the most significant human health improvements globally (Andre et al., 2008). Parents chose not to vaccinate because they are fundamentally concerned about their child's health and well-being. Ironically, it is exactly this reason why parents should vaccinate their children. While vaccine efficacy is not often discussed (due to the ubiquitous nature of vaccines and the now high levels of public health that have been attained due to vaccines), it is imperative to remember that in recent history, and in many parts of the world today, VPDs have led and continue to lead to childhood illness and death far beyond the very few who have suffered an adverse vaccine affect.

NOTES

1. We are operationalizing the term "antivaccinators" to include those who reject all vaccinations and those who purposefully undervaccinate due to concerns over safety. We recognize that there is a spectrum of vaccine skeptics, but for the purpose of this book, we are using the blanket "antivaccinator" label.
2. It should be noted that there is still evidence that undervaccinated children are primarily from low-income homes (Smith et al., 2004), suggesting that income level impacts vaccination choices.
3. Medical exemptions are provided for children who are unable to receive the vaccine due to compromised health and therefore will not be treated here as antivaccination.
4. Due to the use of human cell strains (and some animal cell strains) used in the development of vaccines, some religious groups have questioned the ethics of their use. The Catholic Church, while opposing the use of human cell strains (obtained years ago from two aborted fetuses), has stated support for immunization as the existing cells are not those originally derived directly from the aborted fetus (College of Physicians of Philadelphia, 2017).
5. Thimerosal is still in some flu vaccines, however, but the patient can request thimerosal-free flu vaccines.
6. However, in some studies, education level and science awareness do not correlate due to overriding cognitive bias.

REFERENCES

Anderson, M. "5 Facts about Vaccines in the U.S." Pew Research Center, 2015.

André, F. E. "Vaccinology: Past Achievements, Present Roadblocks and Future Promises." *Vaccine* 21 (2003): 593–95.

Andre, F. E., R. Booy, H. L. Bock, J. Clemens, S. K. Datta, T. J. John, B. W. Lee, S. Lolekha, H. Peltola, T. A. Ruff , M. Santosham, and H. J. Schmitt. "Vaccination Greatly Reduces Disease, Disability, Death and Inequity Worldwide." *Bulletin of the World Health Organization* 86, no. 2 (2008): 140–46.

Baldassare, M., D. Bonner, D. Kordus, and L. Lopez. "Public Statewide Survey: Californians and Their Government." Public Policy Institute of California, 2015.

Blume, S. "Anti-vaccination Movements and Their Interpretations." *Social Science and Medicine* 62, no. 3 (2006): 628–42.

Breslow, J. M., and C. Amico. "The Vaccine War: What Are the Vaccine Exemption Laws in Your State?" *Frontline*, March 24, 2015.

Brunson, E. K. "The Impact of Social Networks on Parents' Vaccination Decisions." *Pediatrics* 131, no. 5 (2013): 1397–404.

Carson, R. *Silent Spring*. New York: Houghton Mifflin, 1962.

Centers for Disease Control and Prevention. "Pertussis Frequently Asked Questions." 2014a.

Centers for Disease Control and Prevention. "U.S. Infant Vaccination Rates High." 2014b.

Centers for Disease Control and Prevention. "Year in Review: Measles Linked to Disneyland." *Public Health Matters Blog*, December 2, 2015.

Centers for Disease Control and Prevention. "History of Smallpox." 2016a.

Centers for Disease Control and Prevention. "Measles Vaccination." 2016b.

Centers for Disease Control and Prevention. "Autism Spectrum Disorder (ASD): Data and Statistics." 2017a.

Centers for Disease Control and Prevention. "Pertussis Frequently Asked Questions." 2017b.

Centers for Disease Control and Prevention. "Possible Side Effects from Vaccines." 2017c. http://www.cdc.gov/vaccines/vac-gen/side-effects.htm.

Christensen, J. "Thimerosal: Everything You Need to Know about This Vaccine Preservative." CNN, February 15, 2017.

College of Physicians of Philadelphia. "Human Cell Strains in Vaccine Development." In *The History of Vaccines*, 2017. https://www.historyofvaccines.org/content/articles/human-cell-strains-vaccine-development.

Conis, E. *Vaccine Nation: America's Changing Relationship with Immunizations*. Chicago: University of Chicago Press, 2015.

DeStefano, F., C. S. Price, and E. S. Weintraub. "Increasing Exposure to Antibody-Stimulating Proteins and Polysaccharides in Vaccines Is Not Associated with Risk of Autism." *Journal of Pediatrics* 163, no. 2 (2013): 561–67.

Fox, S. "The Social Life of Health Information, 2011." Pew Research Center, May 12, 2011.

Freed, G. L., S. J. Clark, A. T. Butchart, D. C. Singer, and M. M. Davis. "Parental Vaccine Safety Concerns in 2009." *Pediatrics* 125 (2010): 654–59.

Funk, C., B. Kennedy, and M. Hefferon. "Vast Majority of Americans Say Benefits of Childhood Vaccines Outweigh Risks." Pew Research Center, February 2, 2017.

Gerber, J. S., and P. A. Offit. "Vaccines and Autism: A Tale of Shifting Hypotheses." *Clinical of Infectious Disease* 48, no. 4 (2009): 456–61.

Goldstein, A. "Tom Price Belongs to a Doctors Group with Unorthodox Views on Government and Health Care." *Washington Post*, February 9, 2017.

Gowda, C., and A. F. Dempsey. "The Rise (and Fall?) of Parental Vaccine Hesitancy." *Human Vaccines and Immunotherapeutics* 9, no. 8 (2013): 1755–62.

Grabenstein, J. D. "What the World's Religions Teach, Applied to Vaccines and Immune Globulins." *Vaccine* 31 (2013): 2011–23.

Gust, D. A., N. Darling, A. Kennedy, and B. Schwartz. "Parents with Doubts about Vaccines: Which Vaccines and Reasons Why." *Pediatrics* 122 (2008): 718–25.

Health Resources and Services Administration. "Data and Statistics." U.S. Department of Health and Human Services, 2017.

Hertz-Picciotto, I., and L. Delwiche. "The Rise in Autism and the Role of Age at Diagnosis." *Epidemiology* 20, no.1 (2009): 84–90.

Hviid, A., M. Stellfeld, J. Wohlfahrt, and M. Melbye. "Association between Thimerosal-Containing Vaccine and Autism." *JAMA* 290, no. 13 (2003): 1763–66.

Ingraham, C. "California's Epidemic of Vaccine Denial, Mapped." *Washington Post*, January 27, 2015.

Jones, A. M., S. B. Omer, R. A. Bednarczyk, N. A. Halsey, L. H. Moulton, and D. A. Salmon. "Parents' Source of Vaccine Information and Impact on Vaccine Attitudes, Beliefs, and Nonmedical Exemptions." *Advances in Preventative Medicine* 2012 (2012): 1–8.

Kata, A. "A Postmodern Pandora's Box: Anti-vaccination Misinformation on the Internet." *Vaccine* 28 (2010): 1709–16.

Kennedy, R. F., Jr. "Mercury and Vaccines. Thimerisol [*sic*], Let the Science Speak." News release, January 27, 2017. http://www.globalresearch.ca/mercury-and-vaccines/5571134.

Kounang, N. "Big Pharma's Big Donations to 2016 Presidential Candidates." CNN, February 11, 2016.

Lam, B. "Vaccines Are Profitable, So What?" *Atlantic*, February 10, 2015. https://www.theatlantic.com/business/archive/2015/02/vaccines-are-profitable-so-what/385214/.

Lehdonvirta, V. "Online Spaces Have Material Culture: Goodbye to Digital Post-Materialism and Hello to Virtual Consumption." *Media, Culture, and Society* 32, no. 5 (2010): 883–89.

Levitz, E. "Trump Health Secretary Says States Should Only Require Vaccines If They Feel Like It." *New York Magazine*, March 16, 2017.

Lewandowsky, S., G. E. Gignac, and K. Oberauer. "The Role of Conspiracist Ideation and Worldviews in Predicting Rejection of Science." *PLoS ONE* 8, no. 10 (2013): e75637.

Lupton, R., and C. Hare. "Conservatives Are More Likely to Believe That Vaccines Cause Autism." *Washington Post*, March 1, 2015.

Lyall, K., R. J. Schmidt, and I. Hertz-Picciotto. "Maternal Lifestyle and Environmental Risk Factors for Autism Spectrum Disorders." *International Journal of Epidemiology* 43, no. 2 (2014): 443–64.

Madsen, K. M., A. Hviid, M. Vestergaard, D. Schendel, J. Wohlfahrt, P. Thorsen, J. Olsen, and M. Melbye. "A Population-Based Study of Measles, Mumps, and Rubella Vaccination and Autism." *New England Journal of Medicine* 347 (2002): 1477–82.

McCright, A. M., and R. E. Dunlap. "Defeating Kyoto: The Conservative Movement's Impact on U.S. Climate Change Policy." *Social Problems* 50, no. 3 (2003): 348–73.

Mergler, M. J., S. B. Omer, W. K. Y. Pan, A. M. Navar-Boggan, W. Orenstein, E. K. Marcuse, J. Taylor, M. P. deHart, T. C. Carter, A. Damico, N. Halsey, and D. A. Salmon. "Association of Vaccine-Related Attitudes and Beliefs between Parents and Health Care Providers." *Vaccine* 31 (2013): 4591–95.

Mnookin, S. *The Panic Virus: The True Story Behind the Vaccine-Autism Controversy*. New York: Simon and Schuster, 2011.

NBC News. "Measles Outbreak Traced to Disneyland Is Declared Over." April 17, 2015. http://www.nbcnews.com/storyline/measles-outbreak/measles-outbreak-traced-disneyland-declared-over-n343686.

PBS. "Refrigerator Mothers: History of Autism Blame." PBS Premiere, July 16, 2002. http://www.pbs.org/pov/refrigeratormothers/fridge/.

Peeples, L. "Anti-vaccine Haven Digs In as Measles Outbreak Hands Science Crusaders and Edge." *Huffington Post*, March 9, 2015.

Pew Research Center. "Americans, Politics, and Science Issues." 2015a.

Pew Research Center. "Public and Scientists' Views on Science and Society." 2015b.

Pew Research Center. "Raising Kids and Running a Household: How Working Parents Share the Load." 2015c.

Price, C. S., W. W. Thompson, B. Goodson, E. S. Weintraub, L. A. Croen, V. L. Hinrichsen, M. Marcy, A. Robertson, E. Eriksen, E. Lewis, P. Bernal, D. Shay, R. L. Davis, and F. DeStefano. "Prenatal and Infant Exposure to Thimerosal from Vaccines and Immunoglobulins and Risk of Autism." *Pediatrics* 126, no. 4 (2010): 656–64.

Rainie, L. "The Online Heath Care Revolution." Pew Research Center, 2000.

Ranji, U., and A. Salganicoff. "Balancing on Shaky Ground: Women, Work and Family Health." Henry J. Kaiser Family Foundation, 2014.

Salmon, D. A., M. Z. Dudley, J. M. Glanz, and S. B. Omer. "Vaccine Hesitancy: Causes, Consequences, and a Call to Action." *American Journal of Preventative Medicine* 49, no. 6S4 (2015): S391–98.

Siders, D. "Jerry Brown Signs California Vaccine Bill." *Sacramento Bee*, June 30, 2015.

Smith, M. J., and C. R. Woods. "On-Time Vaccine Receipt in the First Year Does Not Adversely Affect Neuropsychological Outcomes." *Pediatrics* 125, no. 6 (2010): 1134–41.

Smith, P. J., S. Y. Chu, and L. E. Barker. "Children Who Have Received No Vaccines: Who Are They and Where Do They Live?" *Pediatrics* 114, no. 1 (2004): 187–95.

Steel, B. S., P. List, D. Lach, and B. Shindler. "The Role of Scientists in the Environmental Policy Process: A Case Study from the American West." *Environmental Science and Policy* 7 (2004): 1–13.

Steel, B. S., R. L. Warner, N. P. Lovrich, and J. C. Pierce. "The Inglehart-Flanagan Debate over Postmaterialist Values: Some Evidence from a Canadian-American Case Study." *Political Psychology* 13, no. 1 (1992): 61–77.

Stehr-Green, P., P. Tull, M. Stellfeld, P. Mortenson, and D. Simpson. "Autism and Thimerosal-Containing Vaccines: Lack of Consistent Evidence for and Association." *American Journal of Preventative Medicine* 25, no. 2 (2003): 101–6.

Sun, L. H. "Trump Energizes the Anti-vaccine Movement in Texas." *Washington Post*, February 20, 2017.

Taylor, B., E. Miller, R. Lingam, N. Andrews, A. Simmons, and J. Stowe. "Measles, Mumps, and Rubella Vaccination and Bowel Problems or Developmental Regression in Children with Autism: Population Study." *BMJ* 324, no. 7334 (2002): 393–96.

Taylor, L. E., A. L. Swerdfeger, and G. D. Eslick. "Vaccines Are Not Associated with Autism: An Evidence-Based Meta-analysis of Case-Control and Cohort Studies." *Vaccine* 32, no. 29 (2014): 3623–29.

CHAPTER 4

Climate Change

Debating, doubting, or rejecting the basic scientific facts about climate change in the face of the overwhelming evidence and overwhelming scientific opinion will not change those facts.

—Senator Bernie Sanders

INTRODUCTION

In 2014, the Intergovernmental Panel on Climate Change, an international group formed in 1988 by the United Nations Environmental Programme and the World Meteorological Organization to examine climate change science and to provide information to policy makers, released its *Synthesis Report Summary for Policymakers*. This report clearly outlined what is now considered a scientific fact—that anthropogenic climate change is occurring:

> Warming of the climate system is unequivocal, and since the 1950s, many of the observed changes are unprecedented over decades to millennia. The atmosphere and ocean have warmed, the amounts of snow and ice have diminished, sea level has risen, and the concentrations of greenhouse gases have increased. (Intergovernmental Panel on Climate Change, 2014)

Climate change is not a new phenomenon, and scientific consensus on anthropogenic climate change has occurred only after decades of research and investigation. Unlike other issues discussed in this book with high levels of scientific consensus, climate change is unique in that scientists do not just have a general consensus; there is over 97 percent agreement that climate change is happening and is due primarily to human activities (Cook, Nuccitelli et al., 2013; Cook, Oreskes et al., 2016).

To those in the scientific community, climate change represents a serious challenge to Earth's ecosystem and inhabitants. But climate change policy has been slow to respond to the main culprit of human-induced climate change, GHG emissions. International policies like the Kyoto Protocol of 1997, and most recently the 2015 Paris Agreement, are attempts at garnering international cooperation on GHG mitigation. However, lack of significant action on GHG reduction in time to mitigate for some of the more serious impacts of climate change has now shifted policy discussions to include adaptation and resiliency.

While very few could claim ignorance that climate change is part of an international and national dialogue, some politicians and citizens of the United States are still noncommittal to the idea that climate change is due to human activities. Those unconvinced have been able to stymie significant policy progress on reduction of GHGs, while the scientific community continues to issue an SOS in hopes that policy will respond.

CLIMATE CHANGE SCIENCE AND IMPACTS

Human influence on the climate system is clear, and recent anthropogenic emissions of GHGs are the highest in history. Recent climate changes have had widespread impacts on human and natural systems (Intergovernmental Panel on Climate Change, 2014). Climate change "refers to changes in measures of weather like temperature, precipitation, wind patterns, ocean current, and other indicators that occur over several decades or longer" (Lach, 2014: 98). Humans' impact on climate change was first identified more than 100 years ago by Swedish chemist Svante Arrhenius. Arrhenius's research focused on the effect of coal burning on the atmosphere, and he calculated that "doubling the CO_2 content of the planet's atmosphere would raise its temperature by 2.5 to 4.0 degrees Celsius" (4.5–7.2°F) (McNew, 2014). Further research in the late 1950s determined that the amount of carbon dioxide (CO_2) that could be absorbed by oceans was limited (McNew, 2014), highlighting the problem of excessive CO_2 in the atmosphere and the potential impacts on global temperature rise.

The primary factor of anthropogenic climate change is the emission of GHGs such as CO_2, methane, nitrous oxide, and fluorinated gases. Together, the concentration of GHGs creates a barrier by trapping heat in the atmosphere that drives global warming (known as the greenhouse effect). However, it is CO_2, which is emitted primarily through burning oil, coal, and gas, that accounts for 81 percent of all GHG emissions (Environmental Protection Agency, 2016). CO_2 in the atmosphere has increased over 40 percent since the Industrial Revolution (Melillo, Richmond, and Yohe, 2014), during which time mechanized production required immense energy by burning fossil fuels. Due to a rapidly growing global population and rising living standards around the world, the concentration of CO_2 has accelerated in recent decades, when "if not for human activities, global climate would actually have cooled slightly over the past 50 years" (Melillo, Richmond, and Yohe, 2014).

The increase of GHGs into the atmosphere has further altered Earth's natural ability to sequester CO_2. Sequestration of CO_2 by plants, soils, and the ocean can provide a balance to CO_2 emissions. Unfortunately, increased soil erosion, wildfires, and ocean acidification (outputs of climate change) have created a perverse feedback loop, where Earth's resiliency to absorb and sequester CO_2 is diminishing. Therefore the only option to mitigate for climate change is for humans to reduce GHG emissions, as the planet cannot regulate the output anymore.

As a result of climate change, Earth's average temperature has increased by 1.5°F over the past 100 years, and is anticipated to rise another 0.5 to 8.6°F over the next century (Environmental Protection Agency, 2017). The increase in temperature is already having impacts on Earth such as: an increase in rainfall and floods; more frequent and longer drought conditions and heat waves; higher ocean temperatures and an increase in ocean acidity; the melting of ice caps and rising sea levels; accelerated species loss; an overall increase in extreme weather events including frequency and intensity; and an increase in wildfires and insect infestation. Much of human civilization depends on the health of the planet to provide basic needs. As Earth's system becomes unbalanced, humans will have to adapt to rapidly changing conditions that may stress humans' resiliency to climate change.

DIRECT IMPACTS OF CLIMATE CHANGE

Higher levels of CO_2 in the atmosphere have overtaxed many of Earth's natural systems and, in turn, impacted the human social, cultural, and economic structures that depend on natural resources. For example, currently the ocean absorbs about 30 percent of CO_2 in the atmosphere. The excessive input of CO_2 in the ocean has modified the chemistry of the ocean with a disruptive increase in acidity. Ocean acidification is diminishing coral reefs, disrupting the ability of shellfish to form shells,[1] and impacting the adaptability of marine species (Bennett, 2016).

With over half of the global population residing in coastal regions, people's dependency on the ocean for tourism, fishing, and aquaculture (Creel, 2003) is critical to regional economies and food production. The UN Convention of Biological Diversity conducted an overview of research into economic impacts of ocean acidification, finding that "the global cost of ocean acidification impacts on mollusks and tropical coral reefs is estimated to be over US $1 trillion annually before the end of the century" (Secretariat of the Convention on Biological Diversity, 2014: 97). However, even this estimate may not fully capture the costs of ocean acidification: "The actual costs are likely to be in excess of this figure, particularly when taking account of potentially compounding factors such as overfishing, sedimentation and temperature rise" (Secretariat of the Convention on Biological Diversity, 2014: 97).

But the depletion of the marine environment is just one of an array of resource challenges. Current global water supplies and availability are yet another ominous sign of climate change. Water is the very basis of life, providing a multitude of humans' needs like food, sanitation, and energy. At present, many developing countries are water stressed (more demand than availability), a problem that is only expected to worsen as water availability continues to fluctuate or deteriorate. In the United States, increased temperatures will reduce snowpack and the subsequent availability of water during the spring and summer for agriculture, energy production, wildlife, and humans. While certain areas of the United States will see more rainfall and ensuing flooding, other areas in the United States will witness more drought and water scarcity conditions. In order to meet growing demand for clean water and for other uses like food crops, the United States will have to adapt a sophisticated system of redistribution, storage, and pollution control to maintain adequate, potable water.

Climate change is creating overall warmer conditions; "a warmer atmosphere can hold more moisture, and globally water vapor increases by 7 percent for every degree centigrade of warming" (Clark, 2011). Heavier, more intense rainfall increases the risk of flooding and erosion that will stress existing infrastructure and water quality. Hurricane Katrina in 2005 illustrated the immense toll superstorms can have on lives, infrastructure, and the economy. In all, the hurricane cost an estimated $108 billion, claimed 1,833 lives, and is considered both "the single most catastrophic natural disaster in U.S. history" and the "costliest hurricane in U.S. history" (CNN, 2016). In the oft-cited and dangerously dichotomous environment-versus-the-economy debate, climate change is a game changer. The economic component must incorporate the huge budgetary toll of damage to (among other things) infrastructure, human lives, and property; emergency relief; loss of resources; and economic potential.

Since climate change is a global phenomenon, impacts will resonate worldwide. It is estimated that "if future adaptation mimics past adaptation, unmitigated warming is expected to reshape the global economy by reducing average global incomes roughly 23 percent by 2100 and widening global income inequality" (Burke, Hsiang, and Miguel, 2015: 235). Further, a report by Citigroup found that "the cumulative losses to global GDP from climate change impacts . . . from 2015 to 2060 are estimated at $2 trillion to $72 trillion" globally (Channell et al., 2015).

Loss of natural resources can lead to political and economic instability worldwide. Aside from climate change being a planetary health issue, it also has profound implications for U.S. national security:

> Recent actions and statements by members of Congress, members of the UN Security Council, and retired U.S. military officers have drawn attention to the consequences of climate change, including the destabilizing effects of storms, droughts, and floods. Domestically, the effects of climate change could overwhelm disaster-response capabilities. Internationally, climate change may cause humanitarian disasters, contribute to political violence, and undermine weak governments. (Haass, 2007: v)

The outcomes of climate change illustrate a different, grimmer future for the environment and humans. Under the altered climate conditions, humans will continue to witness more frequent extreme weather events, accelerated species extinction, and ecosystem degradation that will increasingly negatively impact human economic, cultural, and social structures.

A small preview into the human health impact of climate change occurred during the Zika outbreak in 2015. The outbreak, which started in Brazil, corresponded to the hottest year on historical record. The increased heat and humidity in the southern United States provided a conduit for transmission of the disease-carrying mosquitoes to parts of Florida and Texas. Scientists blamed climate change and cautioned that "the Zika epidemic, as well as the related spread of a disease called dengue that is sickening as many as 100 million people a year and killing thousands, should be interpreted as warnings" (Gillis, 2016). As the planet continues to warm, the range of virus-transmitting mosquitoes will increase, making more people vulnerable to viruses once geographically contained (Gillis, 2016).

Opponents of climate change policies often claim that the science is uncertain. In reality, climate change science is quite clear, with abundant, well-documented evidence from the last several decades. A comprehensive report of 300 experts, a 60-member advisory committee, and input from citizens and scientists put it quite simply about climate change science, stating that the "evidence tells an unambiguous story: the planet is warming, and over the last half century, this warming has been driven primarily by human activity" (Melillo, Richmond, and Yohe, 2014).

CLIMATE CHANGE AND THE U.S. PUBLIC

In 1988 Dr. James Hansen, a climatologist and head of the NASA Goddard Institute for Space Studies, testified to a U.S. Senate committee that he was 99 percent sure that climate change was the result of human activities (Shabecoff, 1988). In Dr. Hansen's testimony, he stated:

> Altogether the evidence that the earth is warming by an amount which is too large to be a chance fluctuation and the similarity of the warming to that expected from the greenhouse effect represents a very strong case. In my opinion, that [*sic*] the greenhouse effect has been detected, and it is changing our climate now. (Hearings before the Committee on Energy and Natural Resources, 1988)

Hansen's ominous warning is considered a seminal turning point in the debate over anthropogenic climate change by connecting the burning of fossil fuels to atmospheric changes and making climate change a salient political issue.

But Dr. Hansen's testimony came at a time when shifts in environmental support were occurring. During the 1970s, some of the United States' most significant environmental laws, including the Endangered Species Act, the National Environmental Policy Act, and the Clean Air Act, which passed the Senate "without a single nay vote" (Fuller, 2014), were signed into effect with broad bipartisan support and many under a Republican president. But the environmental successes shared by Congress during the 1970s marked the beginning of the end for bipartisan environmental support. Ironically, just as climate change was becoming a salient issue in the United States, fractures between Democrats and Republicans on environmental issues were becoming visible.

Support for environmental laws required regulation on the industries and businesses contributing to environmental degradation. These same groups being regulated were also part of the Republican base and therefore, beginning with the Reagan administration, "it seemed like the entire GOP was becoming afraid of the issue as businesses and coal companies' complaints grew louder" (Fuller, 2014). As a result, the ongoing division between Democrats and Republicans gradually became more pronounced as Republicans backed away from supporting regulatory environmental policies in favor of free markets and laissez-faire capitalism.

It is perhaps in part due to the political cleavages that environmental concern has, in recent decades, weakened. While in the late 1980s and early 1990s awareness about environmental issues translated to higher levels of concern, that concern has diminished among Americans overall in the last decade (Jones, 2015). Even current environmental concern is constricted by scope and issue priorities. In a 2016 the Pew Research Center survey, 74 percent of Americans said that the "country should do whatever it takes to protect the environment" (Anderson, 2017), but this support is tempered by other priorities, namely, the economy. While 47 percent of Americans have indicated that they want elected officials to "address environmental matters," the vast majority of Americans (75%) rank the economy as more of a priority (Anderson, 2017).

Further, while Americans state broad support for environmental protection, there is an apparent disconnect between environmental concern and concern over (and arguably awareness of) climate change. In 2016, the Pew Research Center study reported that only 48 percent of U.S. adults agreed that the "earth is warming mostly due to human activity." (Funk and Kennedy, 2016). And only 41 percent believe that climate change will affect them personally (Saad and Jones, 2016). Therefore it is not readily apparent that the same individuals who are concerned for the environment (74% mentioned above) share that same level of concern over climate change, perhaps in part because they do not believe they will be directly impacted.

More than 100 years of scientific evidence, and more recently scientific consensus, concludes that anthropogenic climate change is altering Earth's atmosphere, which will cause global temperature rise and as a result dramatically impact the planet. Yet U.S. climate change policy has failed to keep pace with changing climate conditions, which could threaten the ability to mitigate climate change effects in the future. Unlike clean water, clean air, or endangered species, climate change is the epitome of an intractable policy problem. While science tells us how to modify behavior to mitigate climate change, the onus is often on global behaviors that require multinational, multilevel buy-in and action. Perhaps more important, with less than half of Americans concurring that climate change is due to human activity and an even smaller group believing that they will be personally affected by climate change, policy makers lack a clear mandate from the American public, which will potentially have dire impacts on the United States and the world.

Climate Mitigation versus the Economy?

The fossil fuel industry continues to dominate energy production in the United States. It is estimated that the fossil fuel industry stands to lose $33 trillion over the next 25 years as a shift toward renewable energy sources occurs (Ryan, 2016). Does this mean the U.S. economy will suffer? Not necessarily. When the United Kingdom reduced GHG emissions by 8.4 percent, the economy grew by 2.6 percent (William, 2017). Further, a study conducted at Stanford University by Jacobson et al. (2015) finds that the expansion of renewable energy has the potential to power each of the 50 U.S. states by 2050, mitigating climate change, improving air quality, providing jobs, and stabilizing the price and availability of energy. The only barriers that they identify are social and political, "due partly to the fact that most people are unaware of what changes are possible and how they will benefit from them and partly due to the fact that many with a financial interest in the current energy industry resist change" (Jacobson et al., 2015: 2115).

What is particularly vexing about climate change as a public policy issue is the overall lack of awareness (and potentially confidence or denial in science) among the American public about the immediacy of the situation. And with a growing cache of verified evidence and impacts, according to a Pew Research Center study in 2016, still only 48 percent of Americans believe that climate change is the result of human activity, compared to 31 percent who feel it is due to natural causes and 20 percent who feel there is no evidence climate change is occurring (Funk and Kennedy, 2016). In addition, while 67 percent of Americans support climate scientists engaging in policy decisions, only 27 percent agree that scientists have reached consensus on anthropogenic climate change (Funk and Kennedy, 2016).

With strong consensus on anthropogenic climate change and abundant evidence of climate change's current impacts, why are certain segments of the public still lagging behind scientific realities? Unfortunately, disbelief in climate change is more a political reality than a physical reality. For economies that are intricately tied to current energy systems, like the United States, climate change policies require a fundamental shift away from the economic structure (dependent on fossil fuel), which provides enough incentive to believe, support, and publically revere climate-denying scientists (those in the 3% at best) that keep saying everything is fine and that climate change is the biggest hoax ever created. Forget scientific consensus, forget the precautionary principle, and instead wager on the slimmest of margins that things will not turn out as bad as predicted and as mounting evidence suggests. Sound logical?

IDEOLOGY, POSITIVISM, AND POSTMATERIALISM

The United States continues to debate climate change policies, such as the Paris Agreement with the United Nations Framework Convention on Climate Change, which President Obama supported during his presidency but that President Trump now vows to withdraw from. Unless the U.S. public issues a mandate to Congress and the president to combat climate change, the United States may spend another decade without taking substantive action. Ideology has been used as a litmus test for support or opposition to climate policies, but do other variables like positivism and postmaterialism provide insight into individual support for or opposition to climate mitigation?

Ideology: During the 2016 presidential race, Republican candidate Donald Trump consistently cast doubt on climate science, saying that global warming was a "hoax" and that "nobody really knows" if climate change exists. For climate deniers, Trump's words were a welcome relief from a perceived liberal onslaught of climate change worry and demand for mitigation. As Nisbet commented, "The persistent gap in perceptions over the past decade suggests that climate change has joined a short list of issues such as gun control or taxes that define what it means to be a Republican or Democrat" (2009: 14).

Perhaps more than any other individual variable, ideology stands out as a determinant of belief in human-induced climate change. In a Pew research study, only 29 percent of conservatives agreed that climate change is occurring due to human activity compared to 76 percent of liberals (Pew Research Center, 2015). Interestingly, even more conservatives (39%) believe that there is no evidence that climate change is occurring (Pew Research Center, 2015).

However, the ideological division has occurred only in recent decades, potentially offering some insight into how to communicate science to people along the political ideology spectrum. The rise of the environmental movement in the 1960s and 70s ran concurrent with bipartisan support for environmental protection. Considering that many of the United States' seminal environmental regulations and laws were enacted during the Nixon and Ford administrations, early indications of environmental policy suggested a bipartisan appeal. However, President Reagan was decidedly less environmentally focused, beginning an era of Republicans and conservatives distancing themselves from environmental policy support (Dunlap, Xiao, and McCright, 2001).

Reagan, who once famously stated that trees cause more pollution than cars, sought to diminish regulatory authority of environmental regulations and laws. However, many of Reagan's efforts to weaken environmental regulation resulted in public outcry, requiring a switch in tactics (Dunlap, 1987). Arguably, the public backlash "taught conservatives (and industry) that it was more efficacious to question the *need* for environmental regulations by challenging the evidence of environmental degradation, rather than the *goal* of environmental protection" (emphases in the original; Dunlap and McCright, 2011: 146). Thus began an era that continues today of climate skepticism that is driven largely by a conservative agenda to cast doubt about climate facts and projections.

Climate change policies are inherently opposed to conservative ideology and pervasive belief in neoliberalism. Conservative neoliberals "oppose regulation, taxation, and other state policies, which do not serve the short-term corporate bottom line and investor accumulations" (Antonio and Brulle, 2011: 196). Essentially, they believe that the free market should reign, unhindered by government regulation. Climate change policies would have to enact strong regulations to restrict the release of CO_2 and other GHGs into the atmosphere. Perhaps more than any other type of environmental policy, restriction of GHGs will require a transition from current energy use to more renewable, cleaner climate options. For conservatives, these types of restrictions directly challenge the unfettered capitalist system that has allowed a small but powerful minority to profit. Thus as the threat of climate change policies became more of a reality, particularly as scientists became more certain of anthropogenic climate change, climate denialists "have engaged in an escalating assault on climate science and scientists, and in recent years on core scientific practices, institutions, and knowledge" (Dunlap and McCright, 2011: 156) in order to promote climate skepticism.

The extent of denialism among the Republican Party (particularly conservatives) against climate change science at its base is a rejection of anything deemed "liberal." Statements on climate change by conservatives reflect the chilling disconnect from scientific consensus; for instance, Senator Ted Cruz (R-Texas) said he believes that "climate change is a global conspiracy cooked up by liberals who want to institute 'massive government control of the economy, the energy sector and every aspect of our lives' " (Kroll, 2016). Former Republican congressman Bob Inglis says that this denial is consistent with a GOP mind-set and "stuck in a cycle of 'rejectionism,' the total refusal to believe or concede any fact associated with the opposing side, no matter how many experts attest to its veracity" (Kroll, 2016). Economist Paul Krugman says that the Republican Party's climate denial is "extraordinary," writing that:

It's true that conservative parties across the West tend to be less favorable to climate action than parties to their left. But in most countries—actually everywhere except America and Australia—these parties nonetheless support measures to limit emissions. And U.S. Republicans are unique in refusing to accept that there is even a problem. Unfortunately, the extremism of one party in one country has enormous global implications. (Krugman, 2015)

The embedded climate denial, then, appears to be a direct rejection of liberal or Democratic Party principles. Although science almost uniformly confirms the existence of human-caused climate change, conservatives maintain cognitive dissonance, believing instead in invalid and unwarranted conspiracy theories about the liberal agenda.

The success of climate skepticism and denialism cannot be understated. Recent research has found that while scientific consensus on anthropogenic climate change is near 100 percent, the U.S. public is still lagging behind what is at this point scientific fact. Currently, about half of the U.S. population feels that climate change is because of human activity, compared to 31 percent who feel it is naturally occurring and 20 percent who see no solid evidence of climate change (Funk and Kennedy, 2016). A significant minority believes that scientists know that climate change is occurring (33%), that they understand the causes (28%), and that they understand the best way to handle climate change (19%) (Funk and Kennedy, 2016). In addition, only 27 percent understand that there is scientific consensus on human-caused climate change (Funk and Kennedy, 2016). The low public knowledge about climate change and scientific consensus among the American public illustrates the successful messaging from conservative climate skeptics.

Even more nefarious in the climate "debate" is the organization of climate skeptics. Posing "alternative evidence," skeptics have a clear goal:

Environmental skepticism is an elite-driven reaction to global environmentalism, organized by core actors within the conservative movement. Promoting skepticism is a key tactic of the anti-environmental counter-movement . . . designed specifically to undermine the environmental movement's efforts to legitimize its claims via science. Thus, the notion that environmental skeptics are unbiased analysts exposing the myths and scare tactics employed by those they label as practitioners of "junk science" lacks credibility. (Jacques, Dunlap, and Freeman, 2008: 364)

Several studies have examined the relationship between political ideology and belief in human-caused climate change both in the United States and internationally. Findings consistently demonstrate that more conservative individuals are less likely to believe in human-caused climate change (McCright and Dunlap, 2011a) and have a marked decrease in concern over climate change (Whitmarsh, 2011; Zia and Todd, 2010). Indeed, "conservatives and Republicans are more likely to dispute or deny the scientific consensus and the claims of the environmental community, thereby defending the industrial capitalist system" (McCright and Dunlap, 2011b: 180). Alternatively, while conservatives and belief in climate change are negatively associated, liberalism is positively associated with belief in human-caused climate change (McCright and Dunlap, 2011b).

Positivism: Climate policy rests firmly on the scientific agreement that anthropogenic climate change is occurring and that the way to significantly reduce climate impacts is through the reduction of GHGs. This means lessening the United States' dependency on fossil fuels like oil, gas, and coal, which are primary drivers of energy in the United States and abroad. Unlike other environmental issues, which often impact specific groups in the United States, climate change action will require a shift from current fossil fuel dependency to renewable energy sources, something that will affect all Americans. The recognition of the magnitude of this challenge has allowed for the science behind climate change to be diminished and publicly challenged by politicians, the media, and industry.

One of the key arguments against climate change policies (like most environmental policy) is the need to balance our economic priorities against environmental concern. This puts people in an interesting conundrum of balancing the facts of climate change against their worldviews. However, if the science remains a constant, and people identify as believing in science as robust and dependable, then the other factors would have to mitigate around climate change science. This does not appear to be the case; indeed, people's beliefs seem to change in response to priorities.

What is more puzzling is not that people's *priorities* shift with the economic conditions, but that their *beliefs* about basic climate facts and their trust in science also appear to change. The condition of the economy should not alter perception of basic facts or the state of scientific opinion. And yet public opinion trends suggest that people are more likely to "deny" various facts today than they were a few years ago (Scruggs and Benegal, 2012: 508). The notion then that science can provide a clear path to solutions to policy problems is inherently tied to other variables that impact trust and belief in scientific findings.

Worldviews are therefore a critical component in whether an individual will accurately understand climate change science. Political ideology and environmental beliefs are both predictors of support for climate science and distinctly correlated with science belief (e.g., more conservative, less environmentally concerned individuals are less supportive or trusting of climate science with the alternate being true for more liberal, environmentally focused individuals). A study by Whitmarsh captures this relationship finding that:

> Beliefs about climate change are fundamentally linked to existing values and worldviews. More than this, though, the findings show that perceptions of the credibility and meaning of evidence, and of the trustworthiness of communicators of climate change information, are determined by the particular way individuals view the world and the value they attach to different objectives. (2011: 697)

This further suggests that science is a part of a broader individual construct that consists not of facts alone but on how those facts are interpreted through the lens of an individual's worldviews and values. That being said, we would expect that people who hold strong positivistic views would also agree with scientific consensus on climate change, as positivism would be part of their overall worldview.

Postmaterialism: One of the primary cornerstones of the environmental movement is attributed to Inglehart's (1997) postmaterialist theory. Postmaterialism asserts that when material needs are met, people can shift focus to quality-of-life issues. The rise of American hegemony after World War II ran concurrent with a stronger economy and thus a shift in focus from exclusively an acquisition of needs (food, shelter, livable wage) to a broader concept of quality-of-life issues that included human health and well-being. Critical to quality of life is a viable, healthy environment both for human health and wellness and for human enjoyment. In this vein, postmaterialists are more likely to identify as environmentalists and participate in environmental causes through direct action, voting for candidates supporting environmental agendas, and joining environmental groups (Inglehart, 1997).

While Inglehart's postmaterialism theory as applied to environmental concern seemed to mirror the growing economic and social security of the 1960s and '70s, recent research has shown that the relationship between postmaterialism and environmental concern is more nuanced with people in both the developing and the developed world expressing concern for the environment (Franzen and Meyer, 2010; Dunlap and York, 2008). As such, Dunlap and York found that "explanations highlighting the role of the new class and their postmaterialist values may have been useful for explaining the emergence of modern environmentalism in North America and Europe, but are clearly inadequate for explaining the global spread of environmental activism and concern" (2008: 551).

Inglehart himself has addressed these variations as due in part to "objective" conditions that directly impact people in low-income countries, "since those societies with the fewest Postmaterialists tend to have the most severe pollution problems" (1995: 68). Therefore while postmaterialism may serve as an indicator of environmental concern in places like the United States, it may not be a strong indicator in low-income areas where environmentalism has more to do directly with the health and welfare of individuals living in daily contact with pollution and other environmental problems.

In a study examining climate change skepticism and postmaterialism, researchers found that postmaterialism was not a strong predictor of climate skepticism (Tranter and Booth, 2015). Instead, skepticism was more prevalent among those with low trust in government and little concern over environmental issues (Tranter and Booth, 2015). It is possible to hold that postmaterialism can be both a predictor of environmental support as well as an inconsistent predictor as individuals weigh many factors into their beliefs and behaviors. Just as people in low-income countries may not hold postmaterialist values but demonstrate environmental concern, situational factors as well as other personal beliefs and values may play a role in environmental concerns like climate change. Finally, it is important to note that Inglehart's cross-national study of 43 societies found that "nowhere are the Materialists more favorable to environmental protection than the Postmaterialists, but in advanced industrial societies Postmaterialists are more than twice as likely to rank high on support for environmental protection than Materialists" (1995: 65).Thus we would expect that in our research postmaterialists are more likely to believe that anthropogenic climate change is real.

ANALYSES

The West Coast public survey included two questions about climate change that were developed by the Pew Research Center (Funk and Kennedy, 2016). The first question asks people: "From what you have read and heard, is there solid evidence that the average temperature on Earth has been getting warmer over the past few decades?" (See Table 4.1.) If people responded "yes," then they are directed to a second question assessing the cause of global warming with the following responses: "mostly because of human activity such as burning fossil fuel"; "mostly because of natural patterns in the earth's environment; and "don't know." For the West Coast survey, we find that 80.8 percent of respondents

Table 4.1 Public Orientations Concerning Climate Change

From what you have read and heard, is there solid evidence that the average temperature on Earth has been getting warmer over the past few decades?

Percent	
80.8	Yes (N = 1,184)
8.5	No (N = 124)
10.8	Don't know (N = 158)

If yes, do you believe that the earth is getting warmer . . .

Percent	
67.0	Mostly because of human activity such as burning fossil fuels (N = 788)
21.3	Mostly because of natural patterns in the earth's environment (N = 250)
11.7	Don't know (N = 138)

believe the earth is getting warmer, with 8.5 percent responding it is not getting warmer and 10.8 percent responding "don't know." When those respondents answering "yes" as to whether the earth is getting warmer were asked the cause, 67 percent said "human activity," 21.3 percent said "natural patterns," and 11.7 percent said "don't know." These percentages believing the earth is getting warmer and that it is human caused are much higher than what the Pew Research Center has found in the recent national polls (Funk and Kennedy, 2016), but once again the three West Coast states are both politically liberal and environmentally conscious when compared to many other areas of the country, which most likely impacts these results.

According to the literature review above, ideology has been found to be an important and significant predictor of climate change beliefs in the United States and the findings in Table 4.2 reinforce this previous research. Over 90 percent of self-described liberals believe the earth is getting warmer compared to 77.7 percent of moderates and 67.8 percent of conservatives. Only 1.5 percent of liberals do not believe the earth is getting warmer compared to 9.4 percent of moderates and 19.7 percent of conservatives. When asked the cause of the earth getting warmer, there are clear differences between conservatives and liberals. Over 52 percent of conservatives responded that the earth is getting warmer because of natural patterns compared to 31.1 percent of moderates and only 3.4 percent of liberals. Eighty-five percent of liberals believe global warming is human caused compared to 55.3 percent of moderates and 37.9 percent of conservatives. Once again, although California, Oregon,

Table 4.2 Political Ideology and Climate Change Orientations

a. Earth has been getting warmer over the past few decades? [Chi-square = 122.287, $p = .000$]

	Liberal *Percent*	Moderate *Percent*	Conservative *Percent*
Yes	90.4	77.7	67.8
No	1.5	9.4	19.7
Don't know	8.2	12.8	12.4
N =	684	382	395

b. If yes, do you believe that the earth is getting warmer . . . [Chi-square = 296.109, $p = .000$]

	Liberal *Percent*	Moderate *Percent*	Conservative *Percent*
Mostly human caused	85.0	55.3	37.9
Mostly natural patterns	3.4	31.1	52.3
Don't know	11.7	13.7	9.8
N =	618	293	264

and Washington are "blue states," ideological differences concerning climate change are consistent with other national studies.

The next set of analyses examines the impact of postmaterialist values on climate change beliefs. As was discussed previously, postmaterialists may be more likely than materialists to believe the earth is getting warmer (see Table 4.3). Over 82 percent of postmaterialists believe the earth is getting warmer compared to 65.2 percent of materialists. Interestingly, those respondents with "mixed" values were very similar to postmaterialists with 81.1 percent believing the earth has been getting warmer over the past few decades. Materialists were more likely to believe the cause of warming is mostly natural patterns (39.5%), while those respondents with mixed values (22.6%) and postmaterialist values (17.5%) were less likely to believe it is due to natural causes.

Next we turn to the impact of belief in positivism and climate change orientations. As discussed previously, given that there is a "consensus" among climate scientists that the earth is getting warmer and that humans are the main cause, we would expect those with high levels of belief in positivism to have similar views about climate change. The data presented in Table 4.4 show that while those with low, medium, and high levels of belief in positivistic science all agree the earth is getting warmer, those with medium (86.5%) and high (86%) levels of belief in positivism were significantly more likely to believe the earth is getting warmer when

Table 4.3 Postmaterialist Values and Climate Change Orientations

a. Earth has been getting warmer over the past few decades? [Chi-square = 17.243, $p = .000$]

	Postmaterialist *Percent*	Mixed *Percent*	Materialist *Percent*
Yes	82.2	81.1	65.2
No	7.6	8.9	9.1
Don't know	10.2	10.0	25.8
N =	550	850	66

b. If yes, do you believe that the earth is getting warmer . . . [Chi-square = 24.586, $p = .000$]

	Postmaterialist *Percent*	Mixed *Percent*	Materialist *Percent*
Mostly human caused	67.5	68.3	41.9
Mostly natural patterns	17.5	22.6	39.5
Don't know	15.0	9.1	18.6
N =	452	681	43

compared to those with lower levels of belief (70.7%). Only 4.9 percent of respondents with high levels of belief in positivism do not believe in climate change compared to 14.3 percent of low-level belief respondents.

Table 4.4 Positivism Beliefs and Climate Change Orientations

a. Earth has been getting warmer over the past few decades? [Chi-square = 55.837, $p = .000$]

	Low *Percent*	Medium *Percent*	High *Percent*
Yes	70.7	86.5	86.0
No	14.3	5.6	4.9
Don't know	15.1	7.9	9.1
N =	518	482	450

a. If yes, do you believe that the earth is getting warmer . . . [Chi-square = 40.838, $p = .000$]

	Low *Percent*	Medium *Percent*	High *Percent*
Mostly human caused	59.0	64.3	78.1
Mostly natural patterns	23.0	24.1	15.6
Don't know	18.0	11.6	6.3
N =	366	415	384

When examining the reasons why the earth is getting warmer among those respondents who believe in climate change, we find that 78.1 percent of high-level believers in positivism responded that it is human caused compared to 64.3 percent of medium-level believers and 59.0 percent of low-level believers in positivism. Twenty-three percent of respondents with low levels of belief and 24.1 percent of medium-level believers responded that natural patterns are responsible for the earth getting warmer, compared to 15.6 percent of respondents with high levels of belief. Clearly the belief in a positivistic science leads to more people believing in global warming and that warming is due to humans.

The final set of analyses examines the independent effects of ideology, postmaterialist values, and positivism while controlling for age, gender, and education. Because the two questions used to assess climate change beliefs are categorical and not rank-ordered, binary logistic regression is used as the multivariate technique instead of regression (ordinary least squares). This technique requires that the dependent variables be coded as dummy variables, which means for the first model, those respondents who said yes, the earth is getting warmer, were coded as a "1" and all other responses were coded as a "0." For the second model, those respondents who believe global warming is human caused were coded as a "1" and all other responses were coded as a "0."

When examining the results of the logistic regression analyses displayed in Table 4.5 we find that the Chi-square statistic is significant for both models, indicating that each model is a good fit. Both models also correctly classify large percentages of respondents (80.6% and 73.8%, respectively), and the Nagelkerke R^2 is .145 for the Earth Getting Warmer Model and .322 for the Human Caused Model. Ideology was found to be statistically significant for each model with a negative coefficient, meaning that liberals are significantly more likely when compared to conservatives to believe in global warming when controlling for other variables in the model, which is exactly what we found in the bivariate results in Table 4.2. The indicator for postmaterialist values is not statistically significant for either model; however, the indicator for positivistic beliefs does have a significant effect in both models. As was the case with the bivariate results in Table 4.4, those with higher levels of belief in a positivistic science are significantly more likely than those respondents with lower levels of belief to believe that the earth has been getting warmer over the past few decades and that the warming is human caused.

As for the independent impact of age, gender, and education, neither age nor gender had a statistically significant impact in either model. Education does have positive and significant impact for each model,

Table 4.5 Logistic Regression Estimates for Climate Change Orientations[a]

Variable:	**Earth Getting Warmer** *Coefficient (Std. Error)*	**Human Caused** *Coefficient (Std. Error)*
Age	.001 (.004)	.002 (.004)
Gender	–.183 (.145)	–.245 (.128)
Education	.104* (.047)	.230*** (.041)
Ideology	–.699*** (.088)	–11.216*** (.082)
Postmaterialist	.163 (.156)	–.135 (.134)
Positivism	.090*** (.014)	.085*** (.013)
Chi-square =	139.263***	400.596***
Nagelkerke R^2 =	.145	.322
Percent Correctly Classified =	80.6	73.8
N =	1,451	1,451

*$p \leq .05$; **$p \leq .01$; ***$p \leq .001$

[a]Earth Getting Warmer Model: 1 = Earth getting warmer, 0 = else; Human Caused Model 1 = human-caused warming, 0 = else.

however, with those respondents with higher levels of education more likely than those respondents with lower level of education to believe the earth is getting warmer and that it is the result of human activities such as burning fossil fuels.

CONCLUSION

As discussed above, in recent decades there has been growing polarization between the ideological left and the ideological right regarding issues of science, including but not limited to climate change beliefs (McCright and Dunlap, 2011a). Mooney (2005) suggests that the extent of this divide amounts to a partisan "war on science," with conservative Republicans outright denying the general scientific consensus on climate change and liberals being more likely to endorse the science. A further distinction concerns lower levels of trust in science, while endorsing free markets. On issues like climate change, "people who embrace a laissez-faire vision of the free market are less likely to accept that anthropogenic greenhouse gas emissions are warming the planet than people with an egalitarian-communitarian outlook" (Lewandowsky, Gignac, and Oberauer, 2013: e75637).

Conversely, those on the left have been generally more likely to embrace science (Steel, Lach, and Satyal, 2006) and more specifically climate change science (McCright and Dunlap, 2011a, 2011b; Marquart-Pyatt et al., 2014). And according to Anthony Giddens, the issue of climate change "offers the opportunity to renew the case against markets that has for so long been associated with left-of-centre traditions" (2011: 48). Giddens further argues that ideology plays a central role in the climate change debate: "There is . . . a left/right tinge to current climate change debates: those who want to respond to climate change through widespread social reform mostly tend towards the political left; most of the authors who doubt that climate change is caused by human agency, on the other hand, are on the right" (2011: 49). The results presented in this chapter confirm that both ideology and trust in science (i.e., a belief in positivistic science) are indeed related to belief in the scientific consensus on climate change and the causes of that change.

But there may be some hope that messaging is getting through to Americans regardless of ideology or other worldviews. A Gallup Poll conducted in March 2016 reports that 64 percent of Americans are worried either a "great deal" or a "fair amount" about global warming (Saad and Jones, 2016). Central to this concern may be recent efforts to illustrate how climate change impacts people's local communities. The notion of "framing" climate change information to bypass people's inherent biases is extremely important as a tool for issue communication. In a study by Wiest et al., the authors found that:

> In general, our study demonstrates the power of local and benefit frames from projected climate change impacts to influence public opinion on climate change for U.S. residents. At the same time, our research shows that framing effects can vary across individuals with different partisan predispositions. (2015: 195)

This research suggests community framing of the issue can provide a conduit for broader climate change mitigation strategies as it divests people of their inherent predispositions toward the issue and instead focuses on outcome impacts to communities.

This, however, is in stark contrast to political voices that continue to act like the proverbial ostrich with its head in the sand, such as the head of the Environmental Protection Agency Scott Pruitt or legislators from states like North Carolina, Florida, Louisiana, and Tennessee that have banned the use of the term "climate change" or even scientific data that illustrate encroaching sea levels on coastal communities. While collectively Americans may be getting more worried about climate change, many of our political leaders are leading us to the edge of a cliff, pretending we will not fall.

NOTE

1. Currently, the Pacific Northwest oyster business is suffering due to an expansive oyster die-off.

REFERENCES

Anderson, M. "For Earth Day, Here's How Americans View Environmental Issues." Pew Research Center, 2017.

Antonio, R. J., and R. J. Brulle. "The Unbearable Lightness of Politics: Climate Change Denial and Political Polarization." *Sociological Quarterly* 52 (2011): 195–202.

Bennett, J. "Ocean Acidification." Smithsonian National Museum of Natural History, 2016.

Burke, M., S. M. Hsiang, and E. Miguel. "Global Non-linear Effect of Temperature on Economic Production." *Nature* 527 (2015): 235–51.

Channell, J., H. R. Jansen, E. Cȩrmi, E. Rahbari, P. Nguyen, E. L. Morse, E. Prior, A. R. Syme, H. R Jansen, E. Rahbari, E. L. Morse, S. M. Kleinman, and T. Kruger. "Energy Darwinism II: Why a Low Carbon Future Doesn't Have to Cost the Earth." Citi GPS: Global Perspectives and Solutions, August 2015.

Clark, D. "How Will Climate Change Affect Rainfall?" *Guardian*, December 15, 2011.

CNN. "Hurricane Katrina Statistics Fast Facts." August 23, 2016.

Cook, J., D. Nuccitelli, S. A. Green, M. Richardson, B. Winkler, R. Painting, R. Way, P. Jacobs, and A. Skuce. "Quantifying the Consensus on Anthropogenic Global Warming in the Scientific Literature." *Environmental Research Letters* 8, no. 2 (2013): 1–7.

Cook, J., N. Oreskes, P. T. Doran, W. R. L. Anderegg, B. Verheggen, E. W. Maibach, J. S. Carlton, S. Lewandowsky, A. G Skuce, S. A. Green, D. Nuccitelli, P. Jacobs, M. Richardson, B. Winkler, R. Painting, and K. Rice. "Consensus on Consensus: A Synthesis of Consensus Estimates on Human-Caused Global Warming." *Environmental Research Letters* 11, no. 4 (2016): 1–7.

Creel, L. "Ripple Effects: Population and Coastal Regions." Population Reference Bureau, 2003.

Dunlap, R. E. "Polls, Pollution, and Politics Revisited: Public Opinion on the Environment in the Reagan Era." *Environment: Science and Policy for Sustainable Development* 29, no. 6 (1987): 6–37.

Dunlap, R. E., and A. M. McCright. "Organized Climate Change Denial." In *The Oxford Handbook of Climate Change and Society*, edited by J. S. Dryzek, R. B. Norgaard, and D. Schlosberg, 99–115. Oxford: Oxford University Press, 2011.

Dunlap, R. E., C. Xiao, and A. M. McCright. "Politics and Environment in America: Partisan and Ideological Cleavages in Public Support for Environmentalism." *Environmental Politics* 10, no. 4 (2001): 23–48.

Dunlap, R. E., and R. York. "The Globalization of Environmental Concern and the Limits of the Postmaterialist Values Explanation: Evidence from Four Multinational Surveys." *Sociological Quarterly* 49 (2008): 529–63.

Environmental Protection Agency. "Climate Change Impacts." 2016.

Environmental Protection Agency. "Overview of Climate Change Science." 2017.

Franzen, A., and R. Meyer. "Environmental Attitudes in Cross-National Perspective: A Multilevel Analysis of the ISSP 1993 and 2000." *European Sociological Review* 26, no. 2 (2010): 219–34.

Fuller, J. "Environmental Policy Is Partisan. It Wasn't Always." *Washington Post*, June 2, 2014.

Funk, C., and B. Kennedy. "The Politics of Climate: Public Views on Climate Change and Climate Scientists." Pew Research Center, October 4, 2016.

Giddens, A. *The Politics of Climate Change*. 2nd ed. Cambridge, UK: Polity Press, 2011.

Gillis, J. "In Zika Epidemic, a Warning on Climate Change." *New York Times*, February 20, 2016.

Haass, R. N. "Foreword to Climate Change and National Security: An Agenda for Action, by Joshua W. Busby." Council on Foreign Relations, 2007.

Hearings before the Committee on Energy and Natural Resources, Senate, 100th Cong. 40 (1988) (Testimony of Dr. James Hansen). http://climatechange.procon.org/sourcefiles/1988_Hansen_Senate_Testimony.pdf.

Inglehart, R.. *Modernization and Postmodernization: Cultural, Economic, and Political Change in 43 Societies*. Princeton, NJ: Princeton University Press, 1997.

Inglehart, R. "Public Support for Environmental Protection: Objective Problems and Subjective Values in 43 Societies." *PS: Political Science and Politics* 28, no. 1 (1995): 57–72.

Intergovernmental Panel on Climate Change. "Climate Change 2014: Synthesis Report: Summary for Policymakers." 2014.

Jacobson, M. Z., M. A. Delucchi, G. Bazouin, Z. A. F. Bauer, C. C. Heavey, E. Fisher, S. B. Morris, D. J. Y. Pekutowski, T. A. Vencill, and T. W. Yeskoo. "100 Percent Clean and Renewable Wind, Water, and Sunlight (WWS) All-Sector Energy Roadmaps for the 50 United States." *Energy and Environmental Science* 8 (2015): 2093–117.

Jacques, P. J., R. E. Dunlap, and M. Freeman. "The Organization of Denial: Conservative Think Tanks and Environmental Scepticism." *Environmental Politics* 17, no. 3 (2008): 349–85.

Jones, J. M. "In U.S., Concern about Environmental Threats Eases." Gallup, 2015.

Kroll, A. "Why Republicans Still Reject the Science of Global Warming." *Rolling Stone*, November 3, 2016.

Krugman, P. "Republicans' Climate Change Denial Denial." *New York Times*, December 4, 2015.

Lach, D. "Climate Change." In *Science and Politics: An A-to-Z Guide to Issues and Controversies*, edited by B. S. Steel. Thousand Oaks, CA: CQ Press, 2014.

Lewandowsky, S., G. E. Gignac, and K. Oberauer. "The Role of Conspiracist Ideation and Worldviews in Predicting Rejection of Science." *PLoS ONE* 8, no. 10 (2013): e75637.

Marquart-Pyatt, S. T., A. M. McCright, T. Dietz, and R. E. Dunlap. "Politics Eclipses Climate Extremes for Climate Change Perceptions." *Global Environmental Change* 29 (2014): 246–57.

McCright, A. M., and R. E. Dunlap. "Cool Dudes: The Denial of Climate Change among Conservative White Males in the U.S." *Global Environmental Change* 21 (2011a): 1163–72.

McCright, A. M., and R. E. Dunlap. "The Politicization of Climate Change and Polarization in the American Public's Views of Global Warming 2001–2010." *Sociological Quarterly* 52 (2011b): 155–94.

McNew, D. "The Forgotten History of Climate-Change Science." May 13, 2014. http://www.npr.org/sections/13.7/2014/05/13/312128173/the-forgotten-history-of-climate-change-science.

Melillo, J. M., T. C. Richmond, and G. W. Yohe, eds. *Climate Change Impacts in the United States: The Third National Climate Assessment*. Washington, DC: U.S. Global Change Research Program, 2014.

Mooney, C. *The Republican War on Science*. Cambridge, MA: Basic Books, 2005.

Nisbet, M. C. "Communicating Climate Change: Why Frames Matter for Public Engagement." *Environment: Science and Policy for Sustainable Development* 51, no. 2 (2009): 12–23.

Pew Research Center. "Public and Scientists' Views on Science and Society." 2015.

Ryan, J. "Fossil Fuel Industry Risks Losing $33 Trillion to Climate Change." *Bloomberg*, July 11, 2016.

Saad, L., and J. M. Jones. "U.S. Concern about Global Warming at Eight-Year High." Gallup, 2016.

Scruggs, L., and S. Benegal. "Declining Public Concern about Climate Change: Can We Blame the Great Recession?" *Global Environmental Change* 22 (2012): 505–15.

Secretariat of the Convention on Biological Diversity. "An Updated Synthesis of the Impacts of Ocean Acidification on Marine Biodiversity." In *Technical*

Series No. 75, edited by S. Hennige, J. M. Roberts, and P. Williamson, 99. Montreal: Secretariat of the Convention on Biological Diversity, 2014.

Shabecoff, P. "Global Warming Has Begun, Expert Tells Senate." *New York Times*, June 24, 1988.

Steel, B. S., D. Lach, and V. Satyal. "Ideology and Scientific Credibility: Environmental Policy in the American Pacific Northwest." *Public Understanding of Science* 15 (2006): 481–95.

Tranter, B., and K. Booth. "Scepticism in a Changing Climate: A Cross-National Study." *Global Environmental Change* 33 (2015): 154–64.

Whitmarsh, L. "Scepticism and Uncertainty about Climate Change: Dimensions, Determinants and Change over Time." *Global Environmental Change* 21 (2011): 690–700.

Wiest, S. L., L. Raymond, and R. A. Clawson. "Framing, Partisan Predispositions, and Public Opinion on Climate Change." *Global Environmental Change* 31 (2015): 187–98.

William, J. "Renewable Resources: The Impact of Green Energy on the Economy." *business.com*, February 22, 2017.

Zia, A., and A. M. Todd. "Evaluating the Effects of Ideology on Public Understanding of Climate Change Science: How to Improve Communication across Ideological Divides?" *Public Understanding of Science* 19, no. 6 (2010): 743–61.

CHAPTER 5

Teen Sex

Abstinence-only education—the best STD and pregnancy delivery system that politicians have ever devised.

—Rachel Maddow

INTRODUCTION

When the Puritans landed on the coast of Massachusetts Bay, they brought with them a desire to rid themselves of the frivolities of the Church of England. Their goal was to live in a land that exhibited a strict devotion to the Bible and a conformity that was in accordance with God. Of primary importance was piety, particularly as it related to sexual behavior. Believing that the root of all sin was sexual desire, Puritans structured a society that treated sex outside of marriage as a crime, with strict physical and social punishments inflicted on those who disobeyed (Agin, 2010).

The United States' puritanical origins established the moral foundation for the way many Americans think about sex today. In a cross-cultural study examining whether puritanism affects work and sex in the United States, researchers found that "not only devout American Protestants but even non-Protestant and less religious Americans exhibited effects associated with traditional work and sex values" (Uhlmann et al., 2011: 318). In the 2014 U.S. Religious Landscape Study conducted by the Pew Research Center, 83 percent of Americans stated they believed in God (either "absolutely certain" or "fairly certain") (Pew Research Center, 2015), indicating that the United States is still a very religious nation, indeed, the most religious of all advanced countries. It is expressly "because American culture is explicitly religious, [that] even less religious individuals often absorb such values implicitly (Uhlmann et al., 2011: 313). Meaning, the religious values held by the majority of Americans permeate throughout society with the overall effect of imprinting those moral judgments on the United States at large.

While Puritan foundations continue to influence religious beliefs and morals in U.S. society, social realities pertaining to teen sex may be influencing attitudes and acceptability. The average age by which teens now have sex is 17 (Guttmacher Institute, 2016a). While adults' view of the morality behind teen sex still lags behind the reality—between 34 percent and 40 percent of respondents to a Gallup Poll felt it was morally acceptable for teens to engage in sexual activity (Dugan, 2015)—the reality is that teens are having sex whether adults believe it is morally acceptable or not. Teen sex raises important questions pertaining to sexual and reproductive health for teenagers, specifically, avoidance of sexually transmitted disease and unintended pregnancy.

Although the United States is often described as a sexually saturated society, the image of American sexuality is often at odds with the reality. Millennials (those born between 1982 and 1999) are certainly more open-minded about sex but in fact have fewer partners than previous generations (Twenge, Sherman, and Wells, 2015). In addition, teens are now more likely to wait to have sex, and when they do, the majority indicated that they are using contraception (Guttmacher Institute, 2016a). Even so, the rate of teen pregnancy in the United States remains "one of the highest in the developing world" (Guttmacher Institute, 2016a: 3); and half of all new STDs occur in people aged 15 to 24 (Forhan et al., 2009). Among these STD diagnoses, approximately half of new infections each year are due to HPV in people aged 15 to 24 (Guttmacher Institute, 2016a).

In a study conducted by Lindberg, Maddow-Zimet, and Boonstra (2016) examining longitudinal data from the National Survey of Family Growth, they found an overall reduction of formal sex education (through schools, churches, etc.), and that 22 percent of females and 30 about their sexual health. In terms of public policy, having accurate, fact-based comprehensive sex education in addition to access to birth control is central to the reduction of STDs and unintended pregnancies. However, some people contend that the best way to avoid STDs or pregnancy is to avoid sex completely. While this is entirely accurate, it is not reasonable based on the reality that teens have sex and abstinence-only programs have been proven ineffective (Trenholm et al., 2007; Clemmitt, 2010). Further, in some cases, abstinence-only education programs have had the unintended effect of adversely impacting teenagers' sexual health (Sexuality Information and Education Council of the United States, 2009b).

The federal government spends almost $190 million per year for sexual health education programs (Sexuality Information and Education Council of the United States, 2017), yet less than half of high schools (and only about 20% of middle schools) provide comprehensive sex education covering 16 nationally recommended topics on sexual health by the CDC (Demissie et al., 2015). And based on the survey of schools in 48 states in 2014, fewer than 40 percent of schools required sex education for graduation (Demissie et al., 2015). Considering that almost "half of all sexually transmitted infections in the United States occur among those under the age of 25" and that "teens today are less likely than they were a decade ago to say they used a condom the last time they had sex" (CDC, 2015a), the United States has a policy problem on how to effectively educate teens about their sexual health.

According to several public surveys, most U.S. adults feel that comprehensive sex education should be taught in school (National Public Radio, 2004; Eisenberg et al., 2008). In fact, in the 2004 poll conducted by National Public Radio, the Kaiser Family Foundation, and Harvard's Kennedy School of Government, 93 percent of respondents believe that sex education should be taught in schools. However, the debate over sex education regarding what to teach is critical to understanding why there is still an ongoing political battle over federal funding. While vast majorities of people prefer comprehensive sex education that includes abstinence and contraception, and is based on medically accurate information, there is a disparity between support for comprehensive sex education and those who are hesitant to provide access and information for young people to make choices on their sexual and reproductive health. In particular, while comprehensive sex education offers insight into sexual health, access to prevention of unintended pregnancies or STDs such as HPV is still up for debate.

HUMAN PAPILLOMAVIRUS (HPV)

As previously mentioned, STDs disproportionately affect young people. The most common STD for a teen to contract is HPV, with about 35 percent of 14- to 19-year-olds afflicted (Kaiser Family Foundation, 2014). Currently, about 79 million Americans are infected with some form of HPV (CDC, 2017). Although HPV can often go away on its own, in some cases it can cause genital warts or cancer. In order to prevent the spread of HPV, and to help prevent negative health outcomes from contracting the virus, encouraging the use of condoms and getting vaccinated can help diminish the health impacts.

Gardasil was the first vaccination against four of the HPV strains that are the primary strains that cause cancer and genital warts (Cervarix is another vaccine for HPV, targeting two strains of the virus). Prior to Gardasil hitting the market, however, HPV was a rarely discussed STD. The primary marketing strategy for the HPV vaccine touted the vaccine's protection of young girls and women from cervical cancer. Cervical cancer was once the "leading cause of cancer deaths for women in the United States" (CDC, 2016b). However, with the advent of Pap tests, which provided women with a reliable early indicator of cervical cancer, cervical cancer rates dropped.

Yet the vaccine manufacturer Merck resurrected the concern over cervical cancer in its efforts to market Gardasil, focusing on cancer prevention for girls as the primary goal of the vaccine. The rollout of the vaccine made some people skeptical of Merck's intentions. First, Merck stock was in decline when it produced Gardasil, adding more distrust among Americans, who "since the turn of the millennium . . . had shown a rapidly escalating loss of trust in the pharmaceutical industry" (Conis, 2015: 232). Second, people are skeptical of vaccines produced for profit.[1] Merck makes $1.7 billion in sales from Gardasil (Lam, 2015), further increasing doubt as to the real impetus for the vaccine. Third, the vaccine was initially marketed (in addition to a lobbying effort to mandate vaccination) to young girls, generating fears among some conservatives and parents that offering protection from HPV would somehow translate to earlier and/or more casual sex even though "there is no evidence that HPV vaccines increase sexual activity among adolescents" (Gostin, 2011: 1700). Finally, the need for the vaccine versus the potential adverse vaccine reactions has been hotly debated. Many feel that Pap tests are the best indicators of cervical cancer (although invasive, there is no vaccine required) as Pap tests have been very successful in reducing cervical cancer. And just like with other vaccines, there are general worries over safety (see Chapter 3, "Childhood Vaccinations").

Historically, vaccine production has not been a successful way for pharmaceutical companies to make money since they are generally administered only once in a lifetime (including a sequence like MMR) or once a year like the flu vaccine. Disentangling the financial skepticism requires, first, the recognition that vaccines are usually low-profit and, second, that "ten years ago, the financial incentives to produce vaccines were so weak that there was growing concern that pharmaceutical companies were abandoning the vaccine business for selling more-profitable daily drug treatments" (Lam, 2015). And since the manufacturing of vaccines is not done by a government agency or program, in order for vaccines to be researched, produced, and distributed, private pharmaceutical companies are the primary option. If the government did vaccine production, there would be concern over oversight, safety, and government vaccine recommendations. Thus vaccine development has been the domain of private companies, which, by definition, have to make a profit in order to continue to develop and produce both vaccines and other pharmaceutical drugs.

The primary way of contracting HPV is through sexual intercourse. Many parents of 11- and 12-year-olds (the recommended age for the HPV vaccination) are reluctant to get a vaccination for their child that is for an STD. The CDC recommends vaccination at this age as it is hopefully well before kids engage in sex, therefore protecting kids early enough that they are vaccinated prior to first sex. Looking at vaccine hesitancy, however, many who object to the HPV vaccine (aside from antivaccination activists) do so on the grounds that HPV is sexually transmitted and the shot is recommended for children 11 to 12 years old, well before sexually active years.

The idea that by agreeing to the HPV vaccination, parents are somehow tacitly supportive of teen sex is a concern among many who oppose HPV vaccination. Opposition therefore tends to come from people holding more traditional views pertaining to sex, e.g., more conservative (Reiter et al., 2011; Constantine and Jerman, 2007) and more religious individuals (Constantine and Jerman, 2007). It is not surprising then that "so-called 'family values' groups, such as Focus on the Family and the Family Research Council, argued that vaccinating girls against a sexually transmitted infection was tantamount to condoning and encouraging premarital sex" (Conis, 2015: 231).

While there exist high levels of support for the HPV vaccine, many teens in the United States are not vaccinated for HPV. In 2016, the average rate of girls receiving the HPV vaccine was 60 percent; for boys it was 50 percent nationwide (CDC, 2016c). While factors, including cost, temporary discomfort from the vaccination, and skepticism over vaccines, generally may affect vaccination rates, the political debate over teen sexuality may have a greater impact on choosing to vaccinate.

Safety is always an ongoing concern regarding vaccinations. To this end, the HPV vaccine has continued to maintain high safety standards. In 2013, a longitudinal review on safety of both brands of the HPV vaccine (Cervarix and Gardasil) in the United States, Australia, and Japan determined that "data from all sources continue to be reassuring about the safety of both vaccines" (WHO, 2013). Further, "the safety profiles of HPV vaccines have been confirmed—out of clinical trials—by clinical practice and their use worldwide, and they have been included in the immunization schedules in 28 countries" (De Vincenzo et al., 2014: 1005). While some children feel pain at the injection site, brief episodes of dizziness, and other minor and temporary side effects, the HPV vaccine does not represent any unique risks to routine vaccinations.

In terms of efficacy, the relative newness of the vaccine still requires some time to fully ascertain the longevity of the protection the HPV vaccine offers. To date, research on the efficacy of the HPV vaccine shows that "the vaccine continues to be immunogenic and well tolerated up to 9 years following vaccination" (De Vincenzo et al., 2014: 999). The longer the vaccine continues to provide data on efficacy, the more information about long-term HPV protection the vaccine can potentially offer. Since teens are more likely to contract STDs, evidence so far suggests that throughout their teens and early 20s, young people will have protection against HPV.

PLAN B—EMERGENCY CONTRACEPTION

Emergency contraception taken within three to five days after sexual intercourse can significantly reduce unintended pregnancy. Emergency contraception (such as Plan B, Ella, and Next Choice) is currently available both over-the-counter and by prescription. Unlike RU-486 (aka the abortion pill), which terminates existing pregnancies, emergency contraception prevents pregnancy after unprotected sex by preventing the release of an egg, the fertilization of an egg, and/or the implantation of an egg in the uterus (Plan B, 2016). While not to be routinely used for contraception, emergency contraception offers an alternative in the event that there is the possibility of pregnancy after unprotected sex.

Overall, teen pregnancies have been declining since the peak in the 1990s, with evidence demonstrating that improved contraceptive use is the primary driver (Boonstra, 2014). There are a couple of reasons for improved contraception use: comprehensive sex education, including abstinence and correct use of contraception provided formally through school, and the evolution of the Internet as a resource for teens to acquire information about birth control (Boonstra, 2014). Basically, as teens become more informed about their sexual health and have improved access to contraception, they are less likely to become pregnant.

Even with teen pregnancy in sharp decline, it is not surprising that those who do get pregnant do so unintentionally. Almost half of all pregnancies in the United States each year are unintended (Guttmacher Institute, 2016c). The number of unintended pregnancies increases significantly among young women and teens, with 98 percent of pregnancies unintended for teens under 15 (CDC, 2015b), and four out of five pregnancies unintended for women 19 and younger (CDC, 2015b). In some cases, this may be due to failure of contraceptive use or not using contraception at all. In those cases where contraception failed, was not used accurately, or not used at all, emergency contraception can offer a safe way to prevent unintended pregnancy.

In 1999, Plan B, also known as the "morning-after pill," was approved for prescription in the United States. In 2003, the manufacturer of Plan B, Teva Pharmaceutical Industries Ltd., applied to the Food and Drug Administration to make Plan B available over the counter without a prescription (Sifferlin, 2013). This move by Teva set off a political debate about whether the drug should be available for use by minors without parental consent or without a prescription. In 2013, after a protracted political battle between the Food and Drug Administration and reproductive rights and public health organizations, the Obama administration appealed the age limit for Plan B, resulting in the U.S. Court of Appeals for the Second Circuit in Manhattan removing age restrictions for the purchase of Plan B (or generic counterparts).

Removal of age restrictions on emergency contraception met with opposition from people who felt that allowing teens easy access to emergency contraception encourages sexual behavior. Research has found that awareness and education on emergency contraception does not encourage sexual activity in teens (Camp, Wilkerson, and Raine, 2003). Further, some conservative groups argued that Plan B induces abortion, although emergency contraception in fact prevents pregnancy and "if a pregnant woman takes EC, it will not disrupt or harm the pregnancy, nor will it pose any risk to her or the fetus" (Harper et al., 2008: 230). Emergency contraception (with the exception of the copper ring, which is nonhormonal) contains higher levels of the female hormone progestin, preventing conception from taking place; it does not end an existing pregnancy.

It is estimated that between 14 percent and 20 percent of teens have used an emergency contraceptive like Plan B (Kaiser Family Foundation, 2014; Gajanan, 2015). Of those teens who did use emergency contraception, the primary reason was due to unprotected sex (47%) followed by concern that birth control methods failed (34%)

(Daniels, Jones, and Abma, 2013). Since emergency contraception like Plan B is highly effective (approximately 90%) at preventing an unwanted pregnancy, access is crucial for teens who want to avoid pregnancy. One barrier for teens to access Plan B (and other over-the-counter emergency contraception) is cost. With prices ranging from $25 to $50, some teens (especially teens in lower economic conditions)

In 2010, unintended teen pregnancy resulted in a cost of $9.4 billion to U.S. taxpayers "for increased health care and foster care, increased incarceration rates among children of teen parents, and lost tax revenue because of lower educational attainment and income among teen mothers" (CDC, 2016a). Teen mothers are much more likely to drop out of school and therefore experience lower income levels (CDC, 2016a). Further, children of teen mothers are more likely to "have lower school achievement and to drop out of high school, have more health problems, be incarcerated at some time during adolescence, give birth as a teenager, and face unemployment as a young adult" (CDC, 2016a). An often-cited reason to restrict access and availability to teens for emergency contraception is that they are too young to make decisions involving their health and well-being. Considering the lifelong impact of teen pregnancy on both the mother and the child, the balance clearly shifts in favor of removing barriers to emergency contraception for teens.

Another concern about over-the-counter accessibility for emergency contraception for teens is the safety of the drug itself. Plan B has consistently been found to meet over-the-counter safety standards, and it is "nontoxic, nonaddictive, and carries minimal side effects" (Harper et al., 2008: 230). Further, there is no indication of negative drug interactions and no danger of overdose (Harper et al., 2008). Overall, emergency contraception carries less risk than birth control pills, has a proven safety record, and is not connected to any serious adverse health impacts or deaths (Cleland et al., 2014). Aside from minor health impacts similar to menstrual cycles, emergency contraception is a safe, relatively effective means to prevent an unwanted pregnancy for teens.

ABSTINENCE-ONLY EDUCATION

True to the United States' puritanical roots, sex education in public schools has been a push and pull between teaching morals, primarily through abstinence-only education, and teaching comprehensive sex education. In the early 1980s until 2010, the federal government spent over $1.5 billion in abstinence-only-until-marriage programs (Sexuality and Information Council of the United States, 2009a). Abstinence-only education programs promoted a morally-laden view of sex before marriage. For example, the Title V abstinence-only-until-marriage program that was enacted as part of the 1996 welfare reform legislation defined abstinence education as a program that "teaches abstinence from sexual activity outside marriage as the expected standard for all school age children" and "teaches that a mutually faithful monogamous relationship in context of marriage is the expected standard of human sexual activity" (Social Security Administration, 2005). Further, while the term "abstinence" has varied definitions, government policies are "frequently defined in moral terms, using language such as 'chaste' or 'virgin' and framing abstinence as an attitude or a commitment" (Santelli et al., 2006: 73).

Much of the focus on sex education came in response to the rise in teen pregnancy and the HIV and AIDS pandemic in the 1980s. While teen pregnancy rates peaked in 1990 and have been in continued decline since that time (Office of Adolescent Health, 2016), the HIV and AIDS crisis required a public health response so that by the mid-1990s, "every state had passed mandates for AIDS education (sometimes tied to general sex ed and sometimes not)" (Rothman, 2014). Although funding toward abstinence-only education was already in the federal budget, religious conservatives used the public health crisis as a wedge to promote abstinence only. In all accuracy, abstinence-only education proponents argued that the only way to prevent pregnancy and STDs was not having sex. In contrast, the argument for comprehensive sex education was that it teaches abstinence but also utilizes medically accurate information about sexual health including the prevention of STDs and pregnancy through contraceptive use.[2]

While the battle over sex education continues, more information has emerged as to the success of abstinence-only programs versus comprehensive sexual education. In a meta-analysis of prevention programs conducted in 2007, results reveal two significant findings. First, of the abstinence programs (some specifically chosen due to their successes) evaluated, there was not "any strong evidence that any abstinence program delays the initiation of sex, hastens the return to abstinence, or reduces the number of sexual partners" (Kirby, 2007: 15). Conversely, "two-thirds of the 48 comprehensive programs that supported both abstinence and the use of condoms and contraceptives for sexually active teens had positive behavioral affects" (Kirby, 2007: 15). The findings were not unique to Kirby's report. Several studies find that comprehensive sex education results in postponing first sex, more likely to use of condoms and contraception (Mueller et al., 2008; Kohler et al., 2008; Lindberg and Maddow-Zimet, 2012), and having a lower risk for

Tennessee Abstinence Only—Racial and Socioeconomic Implications

Research consistently demonstrates strong support for comprehensive sex education to delay first sex, the use of contraception, and the reduction of pregnancy and STDs. Alternatively, abstinence-only programs have had little to no success in these areas. Memphis, Tennessee, has the dubious distinction of leading the nation in cases of chlamydia and gonorrhea infection (McClain, 2015). Tennessee's abstinence-until-marriage sex education policy has a "requirement that teachers stress abstinence until marriage and allows lawsuits to be brought against teachers who distribute contraception or anything that could be perceived as encouraging experimentation" (McClain, 2015). Legislators in Tennessee further banned any conversation of "gateway" sexual behavior, defined as "non-coital sexual activity such as genital touching" (Taylor, 2012) in order to discourage anything but abstinence before marriage. The results of this legislation are blamed in part for the fact that "girls of all races between the ages of 15 and 19 experience the highest rates of chlamydia in the country" (McClain, 2015). Further, Tennessee is above the national average of teen pregnancies with a rate in 2011 of 58 per 1,000 of teens 15 to 19 years old compared to the national average of 52 per 1,000 (Kost, 2015).

Tennessee, a state deep in the Bible Belt, has taken a hard stance against teen sex, with detrimental results. For a state where it is not uncommon for condoms to be locked up from easy access (McClain, 2015), but that has above-average teen pregnancy rates and the dubious distinction of leading the nation in rates of chlamydia and gonorrhea, there seems to be a clear disconnect between fantasy and reality. While some may believe the ideal is abstinence before marriage, the reality is that teens are not receiving the sexual education necessary to make informed decisions on their sexual health. In this way, Tennessee may be denying teens a basic human right to education and access so they can make responsible, informed choices about their health.

pregnancy (Kohler et al., 2008). Further, a study by Lindberg and Maddow-Zimet (2012) found that comprehensive sex education had a positive impact on partner selection (less age discrepancy, unwanted sex, etc.) and better decision making pertaining to riskier sexual situations (e.g., drinking and drug use).

After years of implementation, "studies overwhelmingly have found little or no evidence that abstinence-only courses change teen sexual behavior in ways that would avert pregnancies or the spread of STDs" (Clemmitt, 2010: 269). Due to lack of substantive evidence supporting abstinence-only education, President Barack Obama in his proposed 2017 budget cut federal funding for all abstinence-only education programs, signifying a political end to the debate over comprehensive sex education and abstinence-only programs. Considering that a poll conducted by National Public Radio, the Kaiser Family Foundation, and Harvard's Kennedy School of Government over a decade before (in 2004) found that only 7 percent of Americans felt that sex education should not be taught in school (National Public Radio, 2014), the 2017 budget cut was overdue.

IDEOLOGY, POSITIVISM, AND POSTMATERIALISM

The issue of teen sex and sexual health is sensitive and fraught with disagreement. Parents do not like to think about their teens as sexual beings, and for many, providing education about sex, supporting vaccinations that prevent STDs, and allowing access to emergency contraception is tantamount to a tacit approval for their kids to have sex. Although research consistently demonstrates the benefits of comprehensive sex education over abstinence-only education, some people reject these findings for their own mores, ideology, and values.

Ideology: In each of the three issues—HPV vaccination, Plan B, and abstinence-only education—much of the policy opposing mandates or restrictions comes from the conservative right (with the notable exception of antivaccinators who also oppose the HPV vaccination). Based on the left-right or liberal-conservative ideological viewpoints, this is not particularly surprising. Conservatives value tradition, particularly religious tradition (Jost, 2006), where "family values" maintain tradition, hierarchy, and control over choices pertaining to sex and sexual health. Some of the early advocates for abstinence-only education came from groups like Focus on the Family, a Christian conservative organization, and the Heritage Foundation, a conservative think-tank. The political power of religious conservatives helped establish the 1981 Adolescent Family Life Act, the first federal funding stream dedicated to abstinence-only education. In 1996, these religious conservatives "helped add provisions for abstinence education to the 1996 Welfare Reform Act" (Rothman, 2014), thereby securing funding for abstinence-only programs by garnering support from Republicans in Congress (Lamb, 2013) until the removal of federal funding for abstinence only under President Obama. Currently, the majority of Evangelicals and Republicans support comprehensive sex education in schools (Peter D. Hart Research Associates, 2008), yet studies conducted by religious conservative groups work further to undermine policy efforts for comprehensive sexual education support. For instance, the National Abstinence Education Association as recently as 2013 released a report offering support for abstinence-based (sexual risk avoidance abstinence-based) programs.[3]

Regardless of whether parents want their teens to have sex, teens are indeed having sex. Peron (2012) noted the irony of conservatives opposing comprehensive sex education, writing:

> The conservative impulse, in other areas, is one of harm reduction. With driving, swimming and firearms they claim logic and common sense. These virtues go out the window when sex is mentioned. Conservatives may speak about risk, but they really mean religion. They are always tempted to use law to impose their religious values. Abstinence-only programs don't exist "for the kids," or because of risks. They exist almost wholly because Republicans . . . are trying to placate a religious minority, even if it means exposing teens to greater harms.

The HPV vaccination assumes the same ideological divisions as abstinence-only education and Plan B. While some antivaccinators are opposed to all vaccinations, with ideological variances, those people who would otherwise vaccinate their children and are opposed to HPV do so on the grounds that providing the HPV vaccine for their child is tacit approval of that child engaging in sex. Republican presidential hopeful Michele Bachmann in the 2011 race heavily criticized Governor Rick Perry's 2007 effort to require the HPV vaccine in Texas. Claiming that the vaccine led to mental retardation (based on an anecdotal story that is medically inaccurate), Bachmann strongly opposed the HPV vaccination. According to a report in the *Guardian*, Bachmann did not oppose the hepatitis B vaccine (which is also sexually transmitted) in Minnesota because "while the Hepatitis B vaccine is uncontroversial, social conservatives such as Bachmann strongly oppose the use of the HPV vaccine as endorsing promiscuity" (Adams, 2011). Conversely, several studies find that liberal parents have a higher acceptability of the HPV vaccination (Constantine and Jerman, 2007: Reiter et al., 2011), which may have to do with ideology, views on both vaccine safety and efficacy, and differing views on sex and sexual health for teens.

Conservatives have a long history of opposition to birth control. The recent Supreme Court case involving Hobby Lobby (*Burwell v. Hobby Lobby*, 2014) refusing to provide contraception like Plan B mobilized evangelicals and others on the religious right to use religious principles as grounds for opposition to contraception. And the outcry over Plan B being available to purchase over the counter without a prescription or an age restriction further entrenched conservative opposition. During the presidential primary campaigns of 2015–16, most of the Republican candidates expressed belief that life begins at conception:

> Virtually every major group opposed to abortion takes the position that life begins at fertilization. Any method of contraception that "may prevent implantation if fertilization does occur," in this view, amounts to the termination of a human life . . . By this logic, a presidential candidate seeking to live up to the standards . . . in the anti-abortion community must then agree that the IUD and morning after pill cause abortions. (Edsall, 2015)

Alternatively, liberals Democrats have long favored access to birth control, as evidenced by President Obama ending the legal battle over age restrictions for Plan B (and other morning-after pills).

Positivism: Comprehensive sex education, the HPV vaccine, and Plan B are all proven highly effective (and safe) at promoting sexual health and safety. In 2007, a study was conducted on the efficacy of abstinence-only Title V–funded programs conducted by Mathematica Policy Research Inc. (on behalf of the U.S. Department of Health and Human Services). The study examined four programs that were thought to be exemplars of the abstinence-only curriculum. Of the four programs that were chosen for potential success, they found no evidence that abstinence-only education was effective at promoting abstinence (Trenholm et al., 2007). Specifically, "findings indicate that youth in the program group were no more likely than the control group to have abstained from sex and, among those who reported having had sex, they had similar number of sexual partners and had initiated sex at the same mean age" (Trenholm et al., 2007: xvii). Alternatively, evidence strongly demonstrates the positive impacts of comprehensive sexual education on delaying first sexual intercourse, increasing contraception use, and reducing pregnancy (Kirby, 2007; Mueller at al., 2008; Kohler et al., 2008; Lindberg and Maddow-Zimet, 2012).

While conservatives have used science to demonstrate that abstinence is the guaranteed method for preventing pregnancy and STDs, this assumes compliance with abstinence. With almost half of high school students reportedly sexually active (Kaiser Family Foundation, 2014), sex education is a pragmatic response to the reality that many teens are sexually active. Further, the reduction in teen pregnancy and teens delaying first sex is positively correlated with comprehensive sex education. Research into teen pregnancy decline finds that 86 percent was the result of improvements in contraceptive use, including increases in the use of individual methods, an increase in the use of multiple methods, and a substantial decline in nonuse (Boonstra, 2014). The remaining 14 percent of the decline could be attributed to a decrease in sexual activity.

While the role of parents in helping to guide and inform their kids' sexual health and choices should not be minimized, as parents are a key influence for teens' information about sex (Albert, 2012), parents want medically accurate information about sex available to their teens. In a 2012 study, 72 percent of adults "believe that teen pregnancy prevention programs that are federally funded should primarily support those programs that have been proven to change behavior related to teen pregnancy" (Albert, 2012: 7). In terms of sex education, the majority of parents want programs with demonstrable results and therefore hold a more positivistic approach to sex education.

The HPV vaccine has demonstrated both longevity of protection through critical teen to young adult sexually active years (De Vincenzo et al., 2014) and safety (WHO, 2013). With rates of HPV cancers on the rise, almost 40,000 new cases each year, vaccination rates in the United States are still low at approximately 42 percent of girls and 28 percent of boys (UC Davis Health, 2017). Although major health organizations like the WHO, National Cancer Institute, and CDC recommend the HPV vaccine based on its low safety risk but high efficacy at preventing cancer and other ill effects from contracting HPV, kids are still undervaccinated in the United States. Because the vaccine rates are significantly lower for HPV than other routine vaccinations, we can assume many people who reject HPV vaccinations otherwise vaccinate their children for things like MMR, chicken pox, tetanus, etc. While rejection of the HPV vaccine may be due to HPV being a sexually transmitted disease, support for the vaccine should be strong among people holding a strong positivistic position.

Like the HPV vaccination, lack of support for Plan B may have less to do with the efficacy and safety of the pill (which is well established) than with its purpose. Those who oppose Plan B do so on the grounds that it is preventing conception. While not an abortion pill, it prevents implantation. Those who hold a strict view about when conception occurs may interpret Plan B as an abortion pill. Again, those holding strong views about science may be less inclined to see the beginning of life at point of fertilization or even implantation. Regardless, conservative opponents continue to raise questions about the drug's "mechanism, calling it a possible abortifacient, although the scientific consensus is that the morning-after pill is a contraceptive that prevents pregnancy by interfering with ovulation and not with implantation of a fertilized egg" (Rabin, 2013).

While science may deem both HPV vaccination and Plan B safe and effective, moral overtones continue to quell the facts.

Postmaterialism: Postmaterialists are much less likely to identify with organized religion when compared to other value types (Inglehart, 1997) and may therefore be less likely to endure the moral conflicts over teen sex and sexual education. Contrary to the conformity of our puritanical roots, postmaterialism moves away from centralized authority and shared norms. Instead there is greater emphasis on quality-of-life priorities (Inglehart, 1971). Whereas much of abstinence-only education derived from conservative, authoritative systems (primarily religious) wherein sexuality and "legitimized sexual expression is always procreative and occurs within the context of an established heterosexual marriage" (Jones, 2011: 136), a more postmaterialist approach would support comprehensive sex education, HPV vaccination, and Plan B as it concedes that teens should be enabled and empowered to make decisions pertinent to their health and well-being.

Sex education, HPV vaccinations, and Plan B clearly involve both genders. However, much of the discussion around these issues disproportionately puts the onus on teen girls to manage their bodies responsibly as they are the ones who can get pregnant (and were initially the focus of the HPV vaccination campaign due to cervical cancer). Postmaterialism is innately tied to feminist beliefs in that postmaterialists "are significantly more likely to endorse feminist beliefs than their materialist counterparts" (Hayes, McAllister, and Studlar, 2000: 436). For example, the HPV vaccination is praised for the benefit to girls' (and boys') sexual health, allowing girls to feel more in control over sexual health choices (Conis, 2015). Consequently, we would expect postmaterialists to be more supportive of comprehensive sex education and approve of the HPV vaccination and Plan B for teens.

ANALYSES

The West Coast public survey contained three questions concerning teenage pregnancy with regard to the HPV vaccine, Plan B (the morning-after pill), and teaching abstinence. The questions included the following with Likert response formats (1 = strongly disagree to 5 = strongly agree): immunizing teens with the HPV vaccination will increase sexual activity; access to Plan B (morning-after pill) will increase sexual activity in teens; and teaching abstinence only in school reduces sexual activity among teens. Responses to these questions can be found in Table 5.1.

Table 5.1 Public Attitudes toward HPV, Plan B, and Abstinence Orientations

Question: How likely are you to agree or disagree with the following statements about? [1 = Strongly disagree to 5 = Strongly agree]

Variable name:		Strongly Disagree *Percent*	Disagree *Percent*	Neutral *Percent*	Agree *Percent*	Strongly Agree *Percent*
HPV	Immunizing teens with the HPV vaccination will increase sexual activity [N = 1,478]	37.6	34.4	21.7	3.0	3.3
Plan B	Access to Plan B (morning-after pill) will increase sexual activity in teens [N = 1,484]	30.9	30.4	18.9	10.1	9.6
Abstinence	Teaching abstinence only in school reduces sexual activity among teens [N = 1,481]	39.6	34.9	14.8	5.5	5.2

Over 60 percent of respondents strongly disagreed and disagreed with each of the three statements. Seventy-two percent of respondents disagreed and strongly disagreed that immunizing teens with the HPV vaccine would increase sexual activity of teens, 21.7 percent responded that they were neutral, and only 6.3 percent agreed and strongly agreed with the statement. In terms of having access to Plan B increasing sexual activity of teens, 61.3 percent disagreed and strongly disagreed with the statement, 18.9 were neutral, and 19.7 percent agreed and strongly agreed with the statement. Out of the three statements, this elicited the highest percentage of respondents in agreement. For the third statement, 74.5 percent strongly disagreed and disagreed, 10.7 percent agreed and strongly agreed, and 14.8 percent were neutral. Clearly there is skepticism about all three claims concerning teenage sexuality among the West Coast respondents. As with the other issues included in this book, perhaps there would be more variation in responses in more conservative areas of the United States.

Correlation coefficients (Tau b) were calculated to assess the relationships between responses to the three statements. Not surprisingly, the

Table 5.2 Correlation Coefficient (Tau b) for HPV, Plan B, and Abstinence Orientations

	Plan B	Abstinence
HPV	.652***	.450***
Plan B		.343***

***$p \leq .001$

coefficients displayed in Table 5.2 show strong relationships between each of the statements with the strongest positive relationship between responses to Plan B and HPV vaccine statements (.652). The smallest of the coefficients was between the statements concerning Plan B and teaching abstinence, but it was still .343. These results indicate that respondents were fairly consistent in their attitudes toward the effect of the HPV vaccine, Plan B, and teaching abstinence on teen sexual activity.

While there is broad disagreement with all three statements by survey respondents, ideology still has a significant impact on people's views as discussed above. The results displayed in Table 5.3 indicate that liberals are significantly more likely to disagree with all three statements when compared to conservatives. Over 83 percent of liberals disagreed that immunizing teens with the HPV vaccine would increase sexual activity compared to 69.6 percent of moderates and 55.1 percent of conservatives. Only 2.8 percent of liberals agreed with this statement compared to 5.2 percent of moderates and 13.6 percent of conservatives.

For the statement concerning access to Plan B increasing teen sexual activity, 80.1 percent of liberals disagreed with the statement compared to 52.8 percent of moderates and 37.2 percent of conservatives. The difference of percentage in agreement is also quite enormous between ideological groups with only 4.9 percent of liberals in agreement compared to 23.3 percent of moderates and 42.2 percent of conservatives. Clearly ideology plays a very strong role in how people view this issue. For the third statement, about teaching abstinence to reduce teen sexual activity, there are also clear differences between ideological groups. While over 64 percent of all three ideological groups disagree with the statement, liberals were most likely to disagree (81.2%) when compared to moderates (72.6%) and conservatives (64.7%). A very small percentage of respondents actually agreed with the statement with the highest percentage for conservatives (15.9%) and the lowest percentage for liberals (7.1%).

Table 5.3 Political Ideology and HPV, Plan B, and Abstinence Orientations

a. Immunizing teens with the HPV vaccination will increase sexual activity [Chi-square = 114.445, p = .000]

	Liberal *Percent*	Moderate *Percent*	Conservative *Percent*
Disagree	83.3	69.6	55.1
Neutral	14.0	25.3	31.3
Agree	2.8	5.2	13.6
N =	687	388	396

b. Access to Plan B (morning-after pill) will increase sexual activity in teens [Chi-square = 273.431, p = .000]

	Liberal *Percent*	Moderate *Percent*	Conservative *Percent*
Disagree	80.1	52.8	37.2
Neutral	14.9	23.8	20.6
Agree	4.9	23.3	42.2
N =	689	390	398

c. Teaching abstinence only in school reduces sexual activity among teens [Chi-square = 38.427, p = .000]

	Liberal *Percent*	Moderate *Percent*	Conservative *Percent*
Disagree	81.2	72.6	64.7
Neutral	11.6	15.9	19.4
Agree	7.1	11.5	15.9
N =	687	390	397

Turning now to the impact of postmaterialist values on HPV vaccine, Plan B, and abstinence orientations, we find that there are some significant differences between value types. As was discussed above in the literature review, materialists have more traditional social norms and values when compared to postmaterialists, who are more likely to value personal freedom and gender equity, and are generally more socially liberal. Therefore we would expect higher levels of disagreement by postmaterialists when compared to other value types. The results displayed in Table 5.4 generally support this line of reasoning.

For the statement concerning immunizing teens with the HPV vaccination contributing to increased sexual activity, 75.6 percent of postmaterialists disagreed with the statement compared to 70.7 percent of

Table 5.4 Postmaterialist Values and HPV, Plan B, and Abstinence Orientations

a. Immunizing teens with the HPV vaccination will increase sexual activity [Chi-square=11.317, p = .023]

	Postmaterialist *Percent*	Mixed *Percent*	Materialist *Percent*
Disagree	75.6	70.7	59.1
Neutral	19.1	22.4	34.8
Agree	5.3	7.0	6.1
N =	549	863	66

b. Access to Plan B (morning-after pill) will increase sexual activity in teens [Chi-square = 48.852, p = .000]

	Postmaterialist *Percent*	Mixed *Percent*	Materialist *Percent*
Disagree	71.6	56.5	37.9
Neutral	13.2	21.6	31.8
Agree	15.2	21.8	30.3
N =	553	865	66

c. Teaching abstinence only in school reduces sexual activity among teens [Chi-square = 11.602, p = .021]

	Postmaterialist *Percent*	Mixed *Percent*	Materialist *Percent*
Disagree	72.5	76.3	68.2
Neutral	18.3	12.4	16.7
Agree	9.2	11.3	15.2
N =	553	862	66

respondents with mixed values and 59.1 percent of materialists. Few of any value type actually agreed with the statement, but over a third of materialists (34.8%) were neutral compared to 22.4 percent of mixed-values types and 19.1 percent of postmaterialists. For the statement that access to Plan B would increase teen sexual activity, we find 71.6 percent of postmaterialists disagreeing with the statement compared to 56.5 percent of mixed-values types and only 37.9 percent of materialists. Slightly over 15 percent of postmaterialists agreed with the statement compared to 21.8 percent of mixed-values types and 30.3 percent of materialists.

For the third statement regarding teaching abstinence to reduce teen sexual activity, the differences are not so prominent between value types.

Mixed- and postmaterialist values types were slightly more likely to disagree with the statement (72.5% and 76.3%, respectively) when compared to materialists (68.2%), and materialists were slightly more likely to agree with the statement (15.2%) when compared to mixed-values types (11.3%) and postmaterialists (9.2%). While we should take caution with the number of materialist-values respondents in the survey, the results are in line with what one would expect for the three statements.

As with the previous chapters, the third set of bivariate analyses examines the impact of positivistic science beliefs. We would expect those with high levels of belief in a positivistic science to more likely disagree with all three statements when compared to respondents with lower levels of belief given that they would be more likely to believe the scientific consensus on these three issues. The results of the analyses are presented in Table 5.5.

Belief in a positivistic science has a significant impact for all three statements concerning the HPV vaccine, Plan B, and abstinence education in the expected direction. Those with high levels of belief in a positivistic science are more likely to disagree that the HPV vaccination will increase teen sexual activity, that access to Plan B will increase teen sexual activity, and that abstinence education will reduce teen sexual activity than those respondents with medium and low levels of belief in positivism. While most low-, medium-, and high-level supporters of positivism all disagree that the HPV vaccination will increase teen sexual activity, those with high levels of belief were 81.6 percent in disagreement compared to 73.1 percent of medium- and 62.3 percent of low-level belief respondents.

As for the statement that access to Plan B will increase teen sexual activity, 68 percent of high-level positivism believers disagreed with the statement compared to 59.7 percent of medium- and 56.6 percent of low-level belief groups. However, there are more respondents in each of the three groups that agreed with this statement when compared to the first statement concerning the HPV vaccine. Twenty-five percent of low-level positivism believers agree that access to Plan B would increase teen sexual activity, and 19 percent of medium- and 14.9 percent of high-level positivism believers also agreed. Nonetheless, those with high-level level positivism beliefs were still the most likely to disagree with the statement.

For the third statement asking respondents if they agree or disagree that teaching abstinence in schools reduces sexual activity of teens, large percentages of low-, medium-, and high-level belief in positivistic science groups are in disagreement. However, like the previous two statements, those with a high level of belief are significantly more likely to disagree

Table 5.5 Positivism Beliefs and HPV, Plan B, and Abstinence Orientations

a. Immunizing teens with the HPV vaccination will increase sexual activity [Chi-square = 50.676, p = .000]

	Low *Percent*	Medium *Percent*	High *Percent*
Disagree	62.3	73.1	81.6
Neutral	29.3	22.3	12.5
Agree	8.4	4.5	5.9
N =	522	484	456

b. Access to Plan B (morning-after pill) will increase sexual activity in teens [Chi-square = 20.510, p = .000]

	Low *Percent*	Medium *Percent*	High *Percent*
Disagree	56.6	59.7	68.0
Neutral	18.4	21.3	17.1
Agree	25.0	19.0	14.9
N =	528	484	456

c. Teaching abstinence only in school reduces sexual activity among teens [Chi-square = 32.306, p = .000]

	Low *Percent*	Medium *Percent*	High *Percent*
Disagree	66.7	77.5	80.9
Neutral	19.0	14.9	10.1
Agree	14.3	7.6	9.0
N =	526	484	456

(80.9%) when compared to medium- (77.5%) and low-level groups (66.7%). The low-level group is most likely to be neutral (19%) and agree with the statement (14.3%) when compared to the medium- and high-level belief groups.

The bivariate survey results indicate that political ideology, belief in positivistic science, and to a lesser extent postmaterialist values do indeed influence West Coast public orientations concerning how teen sexual activity may be influenced by the HPV vaccine, Plan B, and abstinence education. While most respondents were in disagreement that Plan B and the HPV vaccine encourage teen sexual activity, and that teaching abstinence in schools reduces sexual activity, values, ideology, and belief that science is positivistic do influence those orientations.

Table 5.6 Regression Estimates for HPV, Plan B, and Abstinence Orientations

Variable	HPV *Coefficient (Std. Error)*	Plan B *Coefficient (Std. Error)*	Abstinence *Coefficient (Std. Error)*
Age	.004** (.002)	.005** (.002)	.001 (.002)
Gender [1 = female; 0 = male]	.034 (.052)	–.126* (.063)	–.031 (.058)
Education	–.040* (.024)	–.062** (.020)	–.043* (.019)
Ideology	.131*** (.014)	.230*** (.017)	.125*** (.015)
Postmat [1 = postmaterial values; 0 = else]	–.066 (.054)	–.230*** (.066)	.049 (.061)
Positivism	–.035*** (.005)	–.033*** (.006)	–.027*** (.006)
F-test =	33.050***	56.731***	20.154***
Adj. R^2 =	.121	.187	.074
N =	1,438	1,444	1,442

*$p \leq .05$; **$p \leq .01$; ***$p \leq .001$

CONCLUSION

While over 90 percent of Americans feel that sexual education should be taught in schools (National Public Radio, 2004), many American teens are still not receiving full comprehensive sex education. In 2016, Congress funded $176 million for medically accurate, age-appropriate comprehensive sex education (Guttmacher Institute, 2016b). However, only 20 states currently require medically accurate comprehensive sex education, and 35 states and the District of Columbia allow parents to opt out of sex education for their children in public schools (National Conference of State Legislatures, 2016).

By the time teens are 19 years old, 70 percent have had sex (dosomething.org, n.d.). Regardless of personal beliefs about whether or not teens should have sex, teens are having sex. Lack of information about how to prevent pregnancy and STDs has demonstrably proven ineffective, and in some cases has the effect of increasing teen pregnancy and

STD rates. Aside from the financial costs involved in pregnancy and treating STDs, unplanned pregnancy can hinder the academic development of teens and change their trajectory away from future plans to being a parent. The spread of STDs poses a public health issue as well as potentially long-term consequences for teens. Several studies show that providing teens with comprehensive sexual education does not have the effect of tacitly condoning sex. Instead, teens empowered with knowledge are more likely to delay first sex, and when they do have sex are more likely to use contraception including condoms, indicating that they are aware of the prevention of both pregnancy and the spread of STDs (Mueller et al., 2008; Kohler et al., 2008; Lindberg and Maddow-Zimet, 2012).

Further, access to Plan B does not demonstrably correlate with it being used for contraception. Only one-fifth of sexually active teens have ever used Plan B (Kaiser Family Foundation, 2014; Gajanan, 2015), with the primary reason being due to unprotected sex. Plan B is proven safe and effective at preventing pregnancy, providing teens with a responsible option should they have unprotected sex or experience contraceptive failure. While some people may find the idea of Plan B for teens unsettling, teen pregnancy has consequences on both mother and child. Overall, children of teen parents suffer more adverse health effects, are less successful in school, and have higher rates of incarceration (CDC, 2016a); thus there are broad social impacts from teen pregnancy. Further, 35 percent of sexually active teens ages 14 to 19 contract HPV (Kaiser Family Foundation, 2014). HPV can result in cervical cancer (as well as other cancers for both males and females) and have long-term health impacts. While HPV is associated with sexual activity, in reality it is just another vaccination that can prevent and protect teens from illness and disease.

Overall, we found general disagreement that the HPV vaccine, access to Plan B, or comprehensive sexual education increases the sexual activity of teens. Yet differences exist between liberals and conservatives, belief in positivism, and those with postmaterialist versus other value types. Perhaps most notably, conservatives are significantly less likely to disagree that the HPV vaccine, access to Plan B, or comprehensive sexual education increases sexual activity of teens. Teen sex is inherently a political issue, and the United States is still, to some degree, a remnant of our puritanical roots. While sex is ubiquitous in popular media, and prevalent in advertising, movies, and television, Americans struggle with how to balance sexual abundance in our society with messaging to teens on healthy sexual lives. And because sex is inherently a moral issue for many Americans, political debates continue over whether and how to legislate the moral landmine of teen sex.

With strong evidence that teen sexual health is correlated to comprehensive sexual education, HPV vaccination, and access to emergency contraception, the issue is whether Americans want to empower teens to make choices based on knowledge or to impede their ability to prevent pregnancy and the spread of STDs. No amount of government restrictions on sex education or birth control access will prevent teens from having sex. However, public policy can enable teens to make safe choices for their sexual health.

NOTES

1. The HPV vaccine is also expensive in terms of vaccinations, which can be a deterrent for some parents. However, the HPV vaccine is covered under Vaccines for Children, allowing lo- income and/or uninsured people to obtain the vaccine (Conis, 2015).

2. Comprehensive, medically based sex education is also inclusive of topics related to the LGBTQ community, physical abuse and pressure to have sex, and healthy relationships. Abstinence-only education focuses on a narrow view of sex, only heterosexual, and defines marriage as a heterosexual couple, effectively excluding LGBTQ teens.

3. There is little to no evidence supporting these statements by independent researchers. However, research on virginity pledges did find delay of first sex, but when they did have sex they were one-third less likely to use contraception and in some cases had higher rates of STDs (Brückner and Bearman, 2001).

REFERENCES

Adams, R. "Michele Bachmann, the HPV Vaccine and the Republican Landscape." *Guardian*, September 14, 2011.

Agin, D. "Thanksgiving: Puritans, Pilgrims, and Sexual Obsession." *Huffington Post* (2010). March 18, 2010. http://www.huffingtonpost.com/dan-agin/thanksgiving-puritans-pil_b_364724.html.

Albert, B. "With One Voice 2012: America's Adults and Teens Sound Off about Teen Pregnancy." Washington, DC: National Campaign to Prevent Teen and Unplanned Pregnancy, 2012.

Boonstra, H. D. "What Is Behind the Declines in Teen Pregnancy Rates?" *Guttmacher Policy Review*.(September 23, 2014): 15–21.

Brückner, H., and P. S. Bearman. "After the Promise: The STD Consequences of Adolescent Virginity Pledges." *Journal of Adolescent Health* 36, no. 4 (2005): 271–78.

Camp, S. L., D. S. Wilkerson, and T. R. Raine. "The Benefits and Risks of Over-the-Counter Availability of Levonorgestrel Emergency Contraception." *Contraception* 68 (2003): 309–17.

Centers for Disease Control and Prevention. "About Teen Pregnancy: Teen Pregnancy in the United States." News release, April 26, 2016a. https://www.cdc.gov/teenpregnancy/about/.

Centers for Disease Control and Prevention. "Cervical Cancer Statistics." 2016b.

Centers for Disease Control and Prevention. "HPV Vaccine Coverage Maps—Infographic." 2016c.

Centers for Disease Control and Prevention. "Human Papillomavirus (HPV): Genital HPV Infection—Fact Sheet." 2017.

Centers for Disease Control and Prevention. "New Findings from CDC Survey Suggest Too Few Schools Teach Prevention of HIV, STDs, Pregnancy." News release, December 9, 2015a. https://www.cdc.gov/nchhstp/newsroom/2015/nhpc-press-release-schools-teaching-prevention.html.

Centers for Disease Control and Prevention. "Unintended Pregnancy Prevention." 2015b.

Cleland, K., E. G. Raymond, E. Westley, and J. Trussell. "Emergency Contraception Review: Evidence-Based Recommendations for Clinicians." *Clinical Obstetrics and Gynecology* 57, no. 4 (2014): 741–50.

Clemmitt, M. "Teen Pregnancy: Does Comprehensive Sex-Education Reduce Pregnancies?" *CQ Researcher* 20, no. 12 (2010): 265–88.

Conis, E. *Vaccine Nation: America's Changing Relationship with Immunizations*. Chicago: University of Chicago Press, 2015.

Constantine, N. A., and P. Jerman. "Acceptance of Human Papillomavirus Vaccination among Californian Parents of Daughters: A Representative Statewide Analysis." *Journal of Adolescent Health* 40 (2007): 108–15.

Daniels, K., J. Jones, and J. Abma. "Use of Emergency Contraception among Women Aged 15–44: United States, 2006–2010." February 2013. https://www.cdc.gov/nchs/data/databriefs/db112.pdf.

De Vincenzo, R., C. Conte, C. Ricci, G. Scambia, and G. Capelli. "Long-Term Efficacy and Safety of Human Papillomavirus Vaccination." *International Journal of Women's Health* 6 (2014): 999–1010.

Demissie, Z., N. D. Brener, T. McManus, S. L. Shanklin, J. Hawkins, and L. Kann. "School Health Profiles 2014: Characteristics of Health Programs among Secondary Schools." U.S. Department of Health and Human Services, Centers for Disease Control and Prevention, 2015.

dosomething.org. "11 Facts about Sexual Health in Teens in the US." n.d. https://www.dosomething.org/facts/11-facts-about-sexual-health-teens-us.

Dugan, A. "Men, Women Differ on Morals of Sex, Relationships." Gallup, 2015.

Edsall, T. B. "The Republican Conception of Conception." *New York Times*, August 26, 2015.

Eisenberg, M. E., D. H. Bernat, L. H. Bearinger, and M. D. Resnick. "Support for Comprehensive Sexuality Education: Perspectives from Parents of School-Age Youth." *Journal of Adolescent Health* 42 (2008): 352–59.

Forhan, S. E., S. L. Gottlieb, M. R. Sternberg, F. Xu, S. D. Datta, G. M. McQuillan, S. M. Berman, and L. E. Markowitz. "Prevalence of Sexually Transmitted Infections among Female Adolescents Aged 14 to 19 in the United States." *Pediatrics* 124, no. 6 (2009): 1505–12.

Gajanan, M. "Teenage Use of Over-the-Counter Morning-After Pill Doubles in a Decade." *Guardian*, July 25, 2015.

Gostin, L. O. "Mandatory HPV Vaccination and Political Debate." *JAMA* 306, no. 15 (2011): 1699–1700.

Guttmacher Institute. "American Teens' Sexual and Reproductive Health." 2016a.

Guttmacher Institute. "American Teens' Sources of Sexual Health Education." 2016b.

Guttmacher Institute. "Unintended Pregnancy in the United States." 2016c.

Harper, C. C., D. C. Weiss, J. J. Speidel, and T. Raine-Bennett. "Over-the-Counter Access to Emergency Contraception for Teens." *Contraception* 77 (2008): 230–33.

Hayes, B. C., I. McAllister, and D. T. Studlar. "Gender, Postmaterialism, and Feminism in Comparative Perspective." *International Political Science Review* 21, no. 4 (2000): 425–39.

Inglehart, R. *Modernization and Postmodernization: Cultural, Economic, and Political Change in 43 Societies*. Princeton, NJ: Princeton University Press, 1997.

Inglehart, R. "The Silent Revolution in Europe." *American Political Science Review* 4 (1971): 991–1017.

Jones, T. "A Sexuality Education Discourses Framework: Conservative, Liberal, Critical, and Postmodern." *American Journal of Sexuality Education* 6 (2011): 133–75.

Jost, J. T. "The End of the End of Ideology." *American Psychologist* 61, no. 7 (2006): 651–70.

Kaiser Family Foundation. "Sexual Health of Adolescents and Young Adults in the United States." 2014.

Kirby, D. "Emerging Answers 2007: Research Findings on Programs to Reduce Teen Pregnancy and Sexually Transmitted Diseases." National Campaign to Prevent Teen and Unplanned Pregnancy, 2007.

Kohler, P. K., L. E. Manhart, and W. E. Lafferty. "Abstinence-Only and Comprehensive Sex Education and the Initiation of Sexual Activity and Teen Pregnancy." *Journal of Adolescent Health* 42 (2008): 344–51.

Kost, K. "Unintended Pregnancy Rates at the State Level: Estimates for 2010 and Trends since 2002." Guttmacher Institute, 2015.

Lam, B. "Vaccines Are Profitable, So What?" *Atlantic*, February 10, 2015. https://www.theatlantic.com/business/archive/2015/02/vaccines-are-profitable-so-what/385214/.

Lamb, S. "Just the Facts: The Separation of Sex Education from Moral Education." *Educational Theory* 63, no. 5 (2013): 443–60.

Lindberg, L. D., and I. Maddow-Zimet. "Consequences of Sex Education on Teen and Young Adult Sexual Behaviors and Outcomes." *Journal of Adolescent Health* 51 (2012): 332–38.

Lindberg, L. D., I. Maddow-Zimet, and H. Boonstra. "Changes in Adolescents' Receipt of Sex Education 2006–2013." *Journal of Adolescent Health* 58 (2016): 621–27.

McClain, D. "Tennessee's Abstinence-Based Sex-Ed Law Is Especially Bad for Black Students." *Nation*, May 15, 2015.

Mueller, T. E., L. E. Gavin, and A. Kulkarni. "The Association between Sex Education and Youth's Engagement in Sexual Intercourse, Age at First Intercourse, and Birth Control Use at First Sex." *Journal of Adolescent Health* 42 (2008): 89–96.

National Conference of State Legislatures. "State Policies on Sex Education in Schools." 2016.

National Public Radio. "Sex Education in America: An NPR/Kaiser/Kennedy School Poll." February 24, 2004.

Office of Adolescent Health. "Trends in Teen Pregnancy and Childbearing." U.S. Department of Health and Human Services, 2016.

Peron, J. "Sex, Teens and Risk: Conservatives Have It Wrong." *Huffington Post*, April 11, 2012.

Peter D. Hart Research Associates, Inc. "Memorandum: Application of Research Findings", Washington DC: National Women's Law Center and Planned Parenthood Federation of America, Jan.11, 2008. https://nwlc.org/wp-content/uploads/2015/08/RoevsWadeResearch1.11.08.pdf

Pew Research Center. "2014 U.S. Religious Landscape Study." 2015.

Plan B. "How It Works." 2016. http://www.planb.ca/how-it-works.html.

Rabin, R. C. "Morning-After Pill Is Not a Cure-All." Well (blog), *New York Times*, April 8, 2013. https://well.blogs.nytimes.com/2013/04/08/morning-after-pill-is-not-a-cure-all/.

Reiter, P. L., A. McRee, J. A. Kadis, and N. T. Brewer. "HPV Vaccine and Adolescent Males." *Vaccine* 29 (2011): 5595–602.

Rothman, L. "How AIDS Changed the History of Sex Education." *Time*, November 12, 2014.

Santelli, J., M. A. Ott, M. Lyon, J. Rogers, D. Summers, and R. Schleifer. "Abstinence and Abstinence-Only Education: A Review of U.S. Policies and Programs." *Journal of Adolescent Health* 38 (2006): 72–81.

Sexuality Information and Education Council of the United States. "Fact Sheet: Federal Programs Cheat Sheet." 2017.

Sexuality Information and Education Council of the United States. "A History of Federal Funding for Abstinence-Only-Until-Marriage Programs." 2009a.

Sexuality Information and Education Council of the United States. "What the Research Says . . . Abstinence-Only-Until-Marriage Programs." 2009b.

Sifferlin, A. "Timeline: The Battle for Plan B." *Time*, June 11, 2013.

Social Security Administration. "Social Security Act: Title V: Maternal and Child Health Service Block Grant." 2005.

Taylor, M. "Tennessee Sex Ed Bans Mention of "Gateway Sexual Behavior." ABC News, May 13, 2012. http://abcnews.go.com/US/tennessee-governor-passes-controversial-gateway-sexual-behavior-law/story?id=16335600.

Trenholm, C., B. Devaney, K. Fortson, L. Quay, J. Wheeler, and M. Clark. "Impacts of Four Title V, Section 510 Abstinence Education Programs: Final Report." Mathematica Policy Research Institute, April 13, 2007.

Twenge, J. M., R. A. Sherman, and B. E. Wells. "Changes in American Adults' Sexual Behavior and Attitudes, 1972–2012." *Archives of Sexual Behavior* 44, no. 8 (2015): 2273–85.

UC Davis Health. "UC Davis and Nation's Cancer Centers Jointly Endorse Updated HPV Vaccine Recommendations." News release, January 11, 2017. https://www.ucdmc.ucdavis.edu/publish/news/newsroom/11722.

Uhlmann, E. L., T. A. Poehlman, D. Tannenbaum, and J. A. Bargh. "Implicit Puritanism in American Moral Cognition." *Journal of Experimental Social Psychology* 47 (2011): 312–20.

World Health Organization. "World Health Organization Global Advisory Committee on Vaccine Safety, Report of Meeting Held 12–13 June 2013." *Weekly Epidemiological Record* (2013): 309–12.

CHAPTER 6

Stem Cell Research

Embryonic stem cell research has the potential to alleviate so much suffering. Surely, by working together we can harness its life-giving potential.

—Nancy Reagan

INTRODUCTION

In 1998, a scientist at the University of Wisconsin successfully harvested stem cells from unused human embryos from fertility clinics. With the creation of the first human embryonic stem cell (hereafter referred to as ESC) line, opportunities arose for medical advancements that had the potential to cure or treat currently intractable medical issues like Parkinson's disease, heart disease, and cancer. However, concurrent with the optimism over medical possibilities arose ethical concerns concerning the use of human embryos for research.

Stem cell research, while offering a plethora of medical advancements, also remains extremely contentious among those who advocate for people currently living with medical issues and those who feel they must advocate for the lives of the unborn. The first lines of ESCs were created from unused in vitro fertilization (IVF) embryos that would otherwise have been destroyed. Objection over creating and using human ESCs emanates from beliefs about when life begins. For people who believe life begins at fertilization, each of these embryos (from three to five days old) represents a life and therefore deserves the same rights conferred to any human. However, considering that all unused IVF embryos are ultimately destroyed, there exists a logical fallacy pertaining to the use of these embryos for stem cell research (in that we should also see similar opposition to IVF due to the eventual destruction of embryos not used).

There currently exist several ways to obtain stem cells: from adult tissues (bone marrow, brain tissue, tissue from organs, etc.), from embryonic fetal tissue obtained from aborted fetal tissue, and from early-stage human embryos (just fertilized). The primary controversy, and objection to stem cell research, focuses on the one area where the most benefit could derive: human ESCs. ESCs are pluripotent, meaning they have the ability to develop into any of the more than 200 cell types in the human body. This ability offers great medical promise by potentially eliminating the need for pharmaceuticals to manage conditions related to human illness and instead creating a reality "in which ailing organs and tissues might be repaired—not with crude medical devices like insulin pumps and titanium joints but with living, homegrown replacements" (Weiss, 2005: 1).

Alternatively, adult or "somatic" stem cells are currently believed to only differentiate into the cell types from their original organ tissue (National Institutes of Health, 2016). However, with a somatic stem cell line, the cells are derived from adult patients with the intention of developing a stem cell line that could potentially repair tissue from the organ they are derived from, unlike ESC lines, which are malleable enough to become specific types of cells with specialized functions (National Institutes of Health, 2016).

HISTORY AND POLITICS OF ESCs

The discovery of human ESCs came shortly after the birth of the first cloned animal: Dolly the sheep. Dolly was cloned from an adult somatic cell (derived from a mammary gland), which was implanted in a surrogate mother. However, with the announcement of the first ESC lines produced in 1998, the two issues became conflated and soon fears arose regarding human cloning and blurring the line between science and ethics. Scientists marveled at the possibilities ESCs could potentially offer, while opponents feared the overreach of science into the essence of human life.

The conflict over ESC research entered the political realm under the presidency of George W. Bush, when in 2001 he imposed a ban on federal funding supporting the creation of any new stem cell lines, thereby restricting federal research funding to the existing (usable but not robust) 21 lines. For many scientists doing stem cell research, these new restrictions forced separation of federal versus privately funded research, and further constrained collaboration among scientists on stem cell research both domestically and internationally (Murugan, 2009).

Bicameral attempts to resurrect ESC research came in 2005 with the introduction of H.R. 810, formally titled the Stem Cell Research Enhancement Act of 2005. The bill would reestablish federal funding for ESC research and allow for the development of new stem cell lines from discarded embryos from fertility clinics.[1] H.R. 810 received support from both parties, but with heavy support from Democrats. While the bill passed through both houses, President Bush vetoed the bill, with not enough congressional support to overturn the veto. It is perhaps telling that this was the first presidential veto by President Bush in over five years of his presidency. However, the issue was so fraught with moral complexities that, based in large part on ideological and ethical issues, President Bush felt compelled to restrict research on ESCs to appease both his own party and religious supporters.

In 2009, President Obama signed an Executive Order overturning the ban on the use of federal funding for ESC research. While this act opened up hundreds of usable ESC lines, there were still restrictions in place on ESC research due to the Dickey-Wicker amendment of 1996. The Dickey-Wicker amendment (originally a rider on an appropriations bill) restricts federal funds for creating, harming, or destroying human embryos. However, after scientists at the University of Wisconsin created a human stem cell line using private funds, a lawyer (Harriet Rabb) for the Department of Health and Human Services took exception to the definition of embryos under the Dickey-Wicker amendment. Rabb successfully argued that human stem cell lines do not fit the definition of embryos under the Dickey-Wicker amendment, thus opening federally funded research on privately funded and created ESC lines. After Obama revoked the Bush veto on March 9, 2009, the Dickey-Wicker amendment was renewed (in an appropriations bill) two days later on March 11, 2009. However, in 2011 the U.S. Court of Appeals for the District of Columbia Circuit deemed the Dickey-Wicker amendment "ambiguous," thus allowing for the federal funding of research into ESC lines, but upheld the prohibition of destroying embryos for research purposes.

The issue over ESC research squarely lies at the heart of public policy in the United States, spurring the ongoing controversy over "worldviews of the public, pitting religion against the scientific community, universities, and industry" (Nisbet, 2004: 90). Much like abortion, ESC research challenges religious and ethical beliefs regarding when life begins. However, unlike abortion, ESC research is more complex in that it offers the potential for cures for spinal cord injuries, Alzheimer's disease, Parkinson's disease, and other human ailments, with which many people,

Advocating for Stem Cell Research

During George W. Bush's administration, former First Lady Nancy Reagan joined forces with actors Michael J. Fox and Christopher Reeve to criticize the Bush administration for the limitations on ESC research. With former President Ronald Reagan suffering from Alzheimer's disease, Michael J. Fox battling Parkinson's disease, and Christopher Reeve severely disabled due to a spinal injury, each lobbied the Bush administration to support ESC research for the potential benefits it could bring to many in the United States living with debilitating and degenerative conditions.

Consider the case of Parkinson's disease. Currently in the United States, over 1 million people are living with Parkinson's disease, a neurological disorder that slowly impairs movement and progressively worsens over time. There is no cure for Parkinson's disease; however, medications are available to minimize symptoms. Yet these medications often fall short of patient needs due to lack of research on ESCs targeting dopamine neurons, which could potentially effectively mitigate the effects of Parkinson's symptoms for those living with the disease.

For Nancy Reagan, Michael J. Fox, and Christopher Reeve, investment in ESC research bypasses party lines or personal beliefs because it offers hope to the many people living with injuries and disease that impair their quality of life. Yet ESC research remains a highly politicized issue. Even President Obama, while promising to "restore science to its rightful place," did not allow for unrestricted use of embryos or the use of therapeutically cloned embryos. The nuances of ESC research (including but not limited to informed consent, protection of women from scientific exploitation, etc.) make this type of scientific inquiry intrinsically a political and moral issue.

regardless of their religious or ideological worldviews, have direct or indirect experience. Denying research on ESC lines has therefore restricted the possibility for improving the well-being for both individuals afflicted with degenerative or chronic conditions and their families.

Also unique to ESC research is the fact that new lines are created primarily from existing donated embryos (unused from IVF clinics). This raises another tangential issue: since unused embryos are often destroyed, then either way the embryos will be destroyed. However, their use in ESC research means that those possible embryonic lines can potentially be used for medical research promoting advancements that could have an impact on the quality of life for many people living with chronic conditions. Thus ESC research may not necessarily allow for clear ideological or religious support or opposition as the ongoing difficulty in the ESC debate is that "both advocates for and opponents of Federal funding" for ESC research frame the debate in moral terms (Nisbet and Markowitz, 2014).

Many scientists feel that ESC research holds great promise and generally agree that it could potentially unlock the treatments and cures for several debilitating and even fatal medical conditions (Pew Research Center, 2008a). Indeed, few medical advancement opportunities "have captured the imagination of both the scientific community and the public as has the use of stem cells for the repair of damaged tissues" (Fischbach and Fischbach, 2004: 1364). Not surprisingly then, scientists are more likely than the public to support federal funding of ESC research (American Association for the Advancement of Science, 2009).

Vocal opposition to ESC research, and federal funding to support that research, has effectively stymied research efforts to existing ESC lines for almost 20 years. To date, there is little direct evidence of successful use of ESC lines for human health conditions. Rather, "stem cell research holds tremendous promise for medical treatments, but scientists still have much to discover about how stem cells work and their capacity for healing" ("Stem Cells and Medicine," n.d.). In this policy arena, plagued with moral certainty, understanding the divisiveness is helpful but does not necessarily offer policy solutions.

STEM CELLS, IDEOLOGY, AND VALUES

The controversy over the use of stem cells in medical research has been very much influenced by ideological and value differences as alluded to above. As per the theme of this book, the following provides a brief overview of how positivism beliefs, political ideology, and postmaterialist values potentially influence people's orientations toward ESC research.

Ideology: Support or opposition to ESC research is frequently framed in moral terms (e.g., it is either morally wrong or morally the right thing to do). The complexity of ESC research rests on the problem of weighing quality of life with, as some would argue, the right to life. Two separate surveys, conducted by the Pew Research Center and the Gallup Poll in 2013 and 2016, respectively, found that 22 percent to 32 percent of Americans stated that they felt ESC research is "morally wrong" (Pew Research Center, 2013; Gallup, 2016). Further, the Pew Research Center study revealed that approximately 30 percent of Americans feel that federal funding restrictions should be maintained or that no federal funding should be allotted to ESC research.

Among those who feel ESC research is morally wrong, 67 percent identified as conservative (Pew Research Center, 2013). This is not particularly surprising as people in the United States "who regularly attend worship services and hold traditional religious views are much more like to hold conservative political views," while more secular people tend to be more liberal (Pew Research Center, 2008b). The connections between conservative ideology and opposition to ESC research have also been fairly consistent. Data collected between 2002 and 2010 exploring support for ESC research and political ideology found that those who were more conservative were also less supportive of ESC research (Nisbet and Markowitz, 2014).

In a more general context, the debate over ESC research is similar to that of IVF and abortion. Staunch conservatives in both ideology and religious beliefs oppose the creation and subsequent destruction of embryos, which they feel have the potential for life. While this book does not attempt to assert when life begins, it is important to understand that regardless of any benefits that ESC research holds for people suffering from various, often debilitating conditions, people with strong moral objections to the use of embryos for research do not feel the destruction of a potential life merits the benefits for those living with disease and disabilities.

Positivism: While Americans are generally very positive toward scientists, and even welcome their input on policy issues, some are still skeptical of scientific consensus on policy issues that that are contrary to their beliefs. These "scientific pessimists" are less likely to see the benefits of science, including ESC research (Nisbet and Markowitz, 2014). Interestingly, but not surprisingly, scientific pessimists tend to be people who are more conservative and are also less "deferential toward scientific expertise" (Blank and Shaw, 2015: 28). For many, ESC research is a moral, not a scientific issue. The question for these individuals is not whether ESC research has the potential for great medical benefits, but rather whether it should be conducted at all because it destroys embryos in the process. Much like the concern over immunizations, the "facts" of ESC research are irrelevant to the "lives" that are potentially jeopardized.

Alternatively, "scientific optimists" are those who respect science and believe that it has the ability to offer benefits to society, including ESC research (Nisbet and Markowitz, 2014). Again, this is not to suggest that holding a positivistic attitude is exclusive of other value orientations. While dichotomizing positions on science is an easy way to discern support or opposition to science, there are always those who are essentially not interested or disengaged, or those who are conflicted and thus are not always consistent on support for science, particularly on research that challenges inherent views on morality.

Postmaterialism: A defining characteristic of postmaterialist societies is the focus on quality of life. As such, postmaterialists are more likely to support quality of life for themselves and others. Regarding stem cell research, we would anticipate that postmaterialists would support research in this field for two reasons. First, the focus on quality of life implies human health and well-being. People currently living with Parkinson's disease or spinal cord injuries would certainly benefit from any improvement to their conditions that would better their quality of life. Second, postmaterialists are "far more permissive than materialists in their attitudes toward abortion, divorce, extramarital affairs, prostitution, and euthanasia" (Inglehart, 2007). Particularly in the context of stem cell research, postmaterialists' permissive attitudes toward abortion suggest that they would also be supportive of stem cell research.

Postmaterialists are also less traditional in their approach to religious ideals. With secularization more pronounced among postmaterialists (Inglehart, 1997), many may not share the moral conflicts over stem cell research that their materialist counterparts have. Further, many postindustrial, postmaterialist societies now "show declining confidence in churches and falling rates of church attendance and are placing less emphasis on organized religion" (Inglehart, 1997: 328). While organized religion may take a stronger moral stance on ESC research, postmaterialists' more permissive attitudes toward social issues may lessen their dependence on organized religion for guidance on these issues.

ANALYSES

The West Coast public survey contained two statements concerning the use of stem cells in medical research that asked respondents their level of disagreement or agreement for each statement (see Table 6.1). The first statement was: "It is acceptable to use embryonic stem cells for medical research." Slightly over 70 percent of respondents either agreed or strongly agreed with this statement compared to 13 percent disagreeing and strongly disagreeing, and 16.7 percent indicated they were "neutral." Clearly there is strong general public support for stem cell research among the three West Coast states' publics. However, the second statement in the survey elicited a much different response.

The second statement included in the survey asked respondents to agree or disagree with: "Medical research should use stem cells from sources that do NOT involve human embryos." Approximately a third of respondents agreed and strongly agreed with this statement, compared to 35.8 percent disagreeing and strongly disagreeing, and 31.1 percent responding "neutral." While there is strong support in general for the use of stem cells in medical research, there is far less support for using cells

Table 6.1 Public Attitudes toward Stem Cell Research

		Strongly Disagree *Percent*	Disagree *Percent*	Neutral *Percent*	Agree *Percent*	Strongly Agree *Percent*
Stem Cell 1	It is acceptable to use embryonic stem cells for medical research. [N = 1,483]	7.3	5.7	16.7	36.1	34.1
Stem Cell 2	Medical research should use stem cells from sources that do NOT involve human embryos. [N = 1,481]	17.2	18.6	31.1	17.3	15.9

from human embryos. The correlation coefficient (Kendall's Tau b) between responses for both statements was –.384 ($p \leq .000$), indicating that there is indeed variation in how people look at and think about this issue, with many respondents supporting stem cell use in general but not the use of human embryos.

Turning now to the impact of political ideology on stem cell orientations, the survey data displayed in Table 6.2 indicate that there are indeed ideological differences evident among the public. For the first statement—it is acceptable to use ESCs for medical research—a majority of conservatives, moderates, and liberals agreed with the statement. However, liberals were significantly more supportive at 78.4 percent when compared to conservatives at 55.0 percent. Conservatives were most likely to disagree with the statement at 27.9 percent compared to 11.6 percent of moderates and only 5.5 percent of liberals.

When examining the impact of ideology on the second statement—medical research should use stem cells from sources that do not involve human embryos—ideology appears to have an even more pronounced impact with 46.7 percent of liberals disagreeing with the statement and 51.4 percent of conservatives in agreement. Self-described moderates were in between liberals and conservatives with 30.2 percent disagreeing with the statement and 32.7 percent in agreement.

The data displayed in Table 6.3 show disagreement and agreement with the two statements on stem cells and postmaterialist value orientations. Interestingly, the differences between the three value types are not as

Table 6.2 Political Ideology and Stem Cell Orientations

a. It is acceptable to use embryonic stem cells for medical research. [Chi-square = 117.164, $p = .000$]

	Liberal *Percent*	Moderate *Percent*	Conservative *Percent*
Disagree	5.5	11.6	27.9
Neutral	16.1	17.7	17.1
Agree	78.4	70.7	55.0
N =	689	389	398

b. Medical research should use stem cells from sources that do NOT involve human embryos. [Chi-square = 117.168, $p = .000$]

	Liberal *Percent*	Moderate *Percent*	Conservative *Percent*
Disagree	46.7	30.2	21.3
Neutral	30.1	37.1	27.3
Agree	23.1	32.7	51.4
N =	687	388	399

Table 6.3 Postmaterialist Values and Stem Cell Orientations

a. It is acceptable to use embryonic stem cells for medical research. [Chi-square = 20.670, $p = .000$]

	Postmaterialist *Percent*	Mixed *Percent*	Materialist *Percent*
Disagree	15.9	10.9	18.2
Neutral	19.5	14.5	22.7
Agree	64.6	74.7	59.1
N =	553	864	66

b. Medical research should use stem cells from sources that do NOT involve human embryos. [Chi-square = 5.29, p = n.s.]

	Postmaterialist *Percent*	Mixed *Percent*	Materialist *Percent*
Disagree	37.9	34.5	33.3
Neutral	32.1	30.8	25.8
Agree	30.1	34.6	40.9
N =	552	863	66

Table 6.4 Positivism Beliefs and Stem Cell Orientations

a. It is acceptable to use embryonic stem cells for medical research. [Chi-square = 93.805, $p = .000$]

	Low *Percent*	Medium *Percent*	High *Percent*
Disagree	22.7	9.5	5.9
Neutral	20.1	18.0	11.6
Agree	57.2	72.5	82.5
N =	528	484	456

b. Medical research should use stem cells from sources that do NOT involve human embryos. [Chi-square = 21.630, $p = .000$]

	Low *Percent*	Medium *Percent*	High *Percent*
Disagree	29.8	38.6	40.4
Neutral	31.1	28.7	32.9
Agree	39.1	32.6	26.8
N =	524	484	456

pronounced as one might think. Those respondents with "mixed" value orientations were the most likely to agree that ESCs be used in medical research (74.7%) compared to 64.6 percent of postmaterialists and 59.1 percent of materialists. For the second statement concerning the use of human ESCs, there is no statistically significant relationship for postmaterialist values. While 40.9 percent of materialists agree that stem cells from human embryos should not be used compared to 30.1 percent of postmaterialists, the remaining responses among all three value types are not all that different. Clearly more research is needed in this area with perhaps more nuanced multiple indicators of stem cell orientations.

The final set of bivariate analyses concerning stem cell orientations involves belief in scientific positivism. Similar to the results in previous chapters, those respondents who believe in the tenets of positivism are most supportive of the use of stem cells in medical research (see Table 6.4). For those respondents with high levels of belief in positivism, over 82 percent agreed that it is acceptable to use ESCs for medical research and only 5.9 percent disagreed. For those with lower levels of belief in positivism, a majority (57.2%) still agreed that ESC use is acceptable, but 20.1 percent were neutral and 22.7 percent disagreed.

Table 6.5 Regression Estimates for Stem Cell Orientations

Variable	Stem Cell 1 *Coefficient (Std. Error)*	Stem Cell 2 *Coefficient (Std. Error)*
Age	.003 (.002)	–.006** (.002)
Gender	–.243*** (.057)	–.014 (.068)
Education	.097*** (.018)	–.006 (.022)
Ideology	–.182*** (.015)	.187*** (.018)
Postmaterialist	–.049 (.059)	–.076 (.071)
Positivism	.054*** (.005)	–.022*** (.007)
F-test =	66.556***	23.989***
Adjusted R^2 =	.213	.081
N =	1,444	1,441

For the second statement of not using stem cells from human embryos, 26.8 percent of those respondents with high levels of belief in positivism were in agreement compared to 40.4 percent in disagreement. For those with low levels of positivistic belief, 39.1 percent agreed on not using cells from human embryos while 29.8 percent disagreed with the statement. Those with medium levels of belief were more likely to disagree with the statement (38.6%) than agree that human embryo cells should not be used (32.6%). In general, the data support the notion that people with more positivistic views of science are more supportive of stem cells being used for medical research.

The bivariate results indicate that ideology and belief in positivism impact people's orientations toward the use of stem cells in medical research. Postmaterialist values were found to be much less predictive of these orientations. As with the previous chapters, multivariate analyses (ordinary least squares) were used to examine the independent effect of ideology, postmaterialist values, and belief in positivism for both of the stem cell orientation indicators (using the original Likert scale coding of 1 = strongly disagree to 5 = strongly agree) controlling for age, gender, and educational attainment. Table 6.5 provides the results of these analyses.

The F-test for each model is statistically significant, indicating that the models were a good fit. However, the first model for general acceptance of stem cell research in medical research garnered a much higher adjusted R^2 of .213 compared to the model for the use of human embryo stem cells at .081. In terms of the independent impact of value orientations and ideology, the dummy variable for postmaterialism had no significant impact in either model. As the bivariate data presented in Table 6.3 indicated, postmaterialist values did not have a clear and substantial impact on either stem cell statement. However, political ideology and belief in positivism have statistically significant effects in both models with liberals agreeing with the statement that "it is acceptable to use embryonic stem cells in medical research" when compared to more conservative respondents. Similarly, liberals were also more likely than conservatives to disagree with the second statement that "medical research should use stem cells from sources that do NOT involve human embryos."

The results for beliefs in positivism were also as predicted with high levels of belief in positivism associated with agreement with using ESCs for research and disagreement with the statement that medical research should not involve cells from human embryos. As was discussed previously, those who believe in the power of science to be objective, linear, fact finding, and unbiased are more likely to support the efforts of scientists and believe the findings of scientific research.

When examining the results for control variables in each model, women are more likely to agree that stem cells should be used in medical research when compared to men. In addition, the more highly educated are also more likely to support the use of stem cells in research than those with lower levels of education. Age had no statistically significant impact for the use of stem cells in medical research. However, for the second model, only age had a statistically significant impact with younger respondents more likely to disagree with the statement that medical research should not involve stem cells from human embryos when compared to older respondents.

CONCLUSION

Chris Mooney described the scientific consensus on stem cell research as follows:

> Many religious conservatives . . . have asserted that adult stem cells can supplant embryonic ones for research purposes. To the contrary, and despite many insights involving adult stem cells, the scientific consensus remains that the best

research strategy is to pursue both avenues of study simultaneously, because we do not know where research will lead. (2012: 184)

The analysis of public opinion data from the West Coast states of California, Oregon, and Washington revealed wide support for stem cell research in medical research in general, but more divided views concerning the use of stem cells from human embryos that Mooney alludes to in his quote. The analyses presented in this chapter reveal that political ideology does play a role in how people perceive the issue of stem cell research with liberals not only more open to the use of stem cells in general than conservatives but also more open to the use of cells from human embryos. Similarly, those members of the public who have strong beliefs in the ability of science and scientists to further our understanding of how the world works in an unbiased and linear process to find "facts" are also more supportive of stem cell medical research than those who are more skeptical about science and scientists.

The debate over stem cells now has another ethical component pertaining to adult stem cell use. While for some this would provide a backdoor to the moral dilemma, the reprogramming of adult cells into induced pluripotent stem cells (iPS cells) comes with its own set of challenges. Dr. Shinya Yamanaka, who earned a Nobel Prize in Physiology or Medicine in 2012 for his work on iPS cells and the potential for human application. In a recent interview with the *New York Times*, Dr. Yamanaka discussed the ethics of iPS cells, stating:

I think the science has moved too far ahead of the talk of ethical issues. When we succeeded in making iPS cells, we thought, wow, we can now overcome ethical issues of using embryos to make stem cell lines. But soon after, we realized we are making new ethical issues. We can make a human kidney or human pancreas in pigs if human iPS cells are injected into the embryo. But how much can we do those things? It is very controversial. These treatments may help thousands of people. So getting an ethical consensus is extremely important. (Ravven, 2017)

Unlike other contested issues discussed in this book, ESC use and development comes with its own unique set of concerns. When dealing with medical issues like human ESC research and application, U.S. citizens balance quality-of-life issues with morally entrenched concerns about the right to life. What is also unique to ESC research is that as of now, we are discussing the *potential* outcomes of ESC application in humans. Instead of pointing to direct outcomes and arguing over what is, we argue over what could be.

Because ESC research is extremely complex, there is an inherent assumption that if people understood the science, there would be increased support (Nisbet, 2004). As we have discussed in this book, people instinctively rely on heuristic shortcuts to avoid becoming mired in the numerous decisions that they face. Reliance on these shortcuts, however efficient, allows individuals to depend on preformed assumptions, knowledge, and values in guiding their policy preferences. Moral framing of the use of human ESCs delineates support or opposition based on an individual's own ethical compass pertaining to beliefs about when life begins (hence the rights of potential lives) and even more broadly to the role of science and technology to improve the quality of human lives. This cognitive framing alleviates the need to understand the science, which potentially diminishes public understanding of the possible benefits of stem cell use.

NOTE

1. Couples using fertility clinics have few options for unused embryos. They can donate them to research, have them destroyed, or donate them to another couple. Realistically, unused embryos are destroyed, thus meeting the same fate they would if they were used for ESC research.

REFERENCES

American Association for the Advancement of Science. "Pew/AAAS Study: While Public Praises Scientists, Scientists Fault Public, Media." News release, July 9, 2009. https://www.aaas.org/news/pewaaas-study-while-public-praises-scientists-scientists-fault-public-media.

Blank, J. M., and D. Shaw. "Does Partisanship Shape Attitudes toward Science and Public Policy? The Case for Ideology and Religion." *ANNALS of the American Academy of Political and Social Science* 658, no. 1 (2015): 18–35.

Fischbach, G. D., and R. L. Fischbach. "Stem Cells: Science, Policy, and Ethics." *Journal of Clinical Investigation* 114, no. 10 (2004): 1364–70.

Gallup. "Stem Cell Research." In *In Depth: Topics A to Z*. 2016.

Inglehart, R. *Modernization and Postmodernization: Cultural, Economic, and Political Change in 43 Societies*. Princeton, NJ: Princeton University Press, 1997.

Inglehart, R. F. "Postmaterialism." In *Encyclopedia Britannica*. 2007.

Mooney, C. *The Republican Brain: The Science of Why They Deny Science—and Reality*. New York: Wiley, 2012.

Murugan, V. "Embryonic Stem Cell Research: A Decade of Debate from Bush to Obama." *Yale Journal of Biology and Medicine* 82, no. 3 (2009): 101–3.

National Institutes of Health. "Stem Cell Information." 2016.

Nisbet, M. C. "The Competition For Worldviews: Values, Information, and Public Support for Stem Cell Research." *International Journal of Public Opinion* 17, no. 1 (2004): 90–112.

Nisbet, M., and E. M. Markowitz. "Understanding Public Opinion in Debates over Biomedical Research: Looking Beyond Political Partisanship to Focus on Beliefs about Science and Society." *PLoS ONE* 9, no. 2 (2014): 1–12.

Pew Research Center. "Abortion Viewed in Moral Terms: Fewer See Stem Cell Research and IVF as Moral Issues." August 15, 2013.

Pew Research Center. "The Case for Embryonic Stem Cell Research: An Interview with Jonathan Moreno." D. Masci, interviewer. *Religion & Public Life*, July 17, 2008a.

Pew Research Center. "U.S. Religious Landscape Survey: Religious Beliefs and Practices." *Religion & Public Life*, June 1, 2008b.

Ravven, W. "The Stem-Cell Revolution Is Coming—Slowly." *New York Times*, January 16, 2017.

"Stem Cells and Medicine." A Closer Look at Stem Cells, n.d. http://www.closerlookatstemcells.org/stem-cells-and-medicine/.

Weiss, R. "The Power to Divide." *National Geographic*, July 2005.

CHAPTER 7

Conclusion: What Is to Be Done?

Science policy and science politics must coexist in an environment that allows, even encourages, creativity. Independent and critical thinking contributes to the dynamic dialogue; muzzling those whose views are at odds with the majority party or distorting evidence to fit one's point of view is not only bad science, but also bad politics.[1]

INTRODUCTION

As discussed in Chapter 1, across the United States and in other countries many citizens, interest groups, and policy makers have called for *evidence-based and/or science-based* public policy. Underlying this sentiment is faith that scientists and the scientific information they provide can improve the quality of complex public policy decisions. This is an outgrowth of the philosophy of "logical positivism." The common assumption is that where science is relevant to policy issues, dispassionate and "neutral" scientists can and should facilitate the development of public policy by providing scientific insight to the public and policy makers. As the National Research Council's report *Using Science as Evidence in Public Policy* argues, this has been a growing demand due to accountability, resource issues, and the quality of the policy design:

Using science in public policy is on the nation's agenda. One reason is the growing demand for performance measures and enhanced accountability in federal agencies and not-for-profit organizations. Another is the call for evidence-based policy and practice, part of a broader focus on data-driven decision-making across government agencies. (National Research Council, 2012, 21)

While a good many scientists agree that science is not value-free and that alternative sources of information beyond those produced by scientists can lead to the development of valid judgments about many policy issues (Steel et al., 2004), most would agree with Levien (1979) that science and scientists should play a role in public policy. Levien argues that there are three principal ways that science and technology can contribute to the effective management of environmental problems. First, scientists can help provide citizens and decision makers a "common understanding" of the key dimensions of the policy problems being discussed. Second, science can then "describe and invent options for the solution" of the policy problem. And third, science can contribute to the resolution of policy problems by estimating "the consequences of proposed solutions" (1979: 48). However, as the case studies in this book have demonstrated, values and ideology can often influence which scientists and scientific information should be used and which should be ignored.

This concluding chapter will first discuss potentially inherent trade-offs between science, scientific experts, and democracy—often called the "democracy versus technocracy quandary"—and then review the major findings of this book concerning how political ideology, values, and beliefs influence public orientations toward genetically modified organisms, immunizations, climate change, teen sex, and stem cells. In the concluding section of the chapter, various approaches will be discussed on how to reconcile resolve the democracy-versus-technocracy quandary and build trust in science and scientists that transcends ideology and values.

THE DEMOCRACY-VERSUS-TECHNOCRACY QUANDARY

As an advanced industrial democracy, the role of science and scientists in the policy process has posed a "quandary" over the proper degree of involvement and influence science and scientists should have in policy formation. This problem has been referred to as the "democracy versus technocracy quandary," or the "democracy versus science quandary" (Pierce and Lovrich, 2014). As the survey data presented in Chapter 1 revealed, there is much variation among the public concerning the level of engagement that scientists should play in the policy process. About a fifth of West Coast survey respondents wanted scientists to play a minimalist role while over a fourth of respondents thought that scientists should be advocates for specific policies. Almost half of respondents preferred that scientists should either interpret results for policy makers or help policy makers integrate science into policy. Of course, as we have seen in the case study chapters, the public may well prefer different roles for science and scientists depending on the specific policy issue and their ideological and value orientations, further complicating this quandary.

As was further discussed in Chapter 1, there has been a decline in public trust in science in recent years. After examining Pew Research Center survey data, Mark Lynas commented in a *Washington Post* editorial: "Even as science makes unparalleled advances in genomics to oceanography, science deniers are on the march—and they're winning hearts and minds more successfully than the academic experts" (2015). This trend has been concurrent with a similar noteworthy decline in public trust of government. The Pew Research Center has described this similar decline as follows: "Fewer than three-in-ten Americans have expressed trust in the federal government in every major national poll conducted since July 2007–the longest period of low trust in government in more than 50 years" (Pew Research Center, 2015).

Along with this diminished trust in scientists and government officials have come forceful demands for increasing citizen involvement in governance (Inglehart, 1997; Vigoda, 2002). The concern that arises in this context is the need for scientific expertise (technocracy) to develop appropriate options in complex areas of public policy, while simultaneously meeting the demand for the enhancement of direct citizen participation (democracy), resulting in these dual roles coming into direct conflict (Pierce and Lovrich, 2014). On the other hand, excessive democracy in the form of the direct involvement of overly ideological and ill-informed citizens in policy making may relegate scientific information to such a peripheral role that complex problems will be inadequately addressed by the adoption of "political" solutions. As Lynas (2015) described this scenario:

> Effective governance in a democratic society depends on voters being able to make choices based on accurate information. If the voices of scientific experts continue to be drowned out by those of ideologues, whether from left or right, America risks moving farther away from the Enlightenment values on which the republic was founded. Such a shift would harm everyone—whether or not they believe the Earth is warming.

On the one hand, too much emphasis on science and scientists as policy makers risks the erosion of democracy and the progressive incapacitation of the citizenry. A 2011 entry in *Technocrats and Democracy*, a blog in the *Economist*, outlines this particular danger:

> Almost by definition, technocrats command respect rather than popularity: they tend especially to drive the far left and right further to the extremes. And at the

moment, the only politicians who are unquestionably thriving are those outside the mainstream already. . . . Technocrats may be good at saying how much pain a country must endure, how to make its debt level sustainable or how to solve a financial crisis. But they are not so good at working out how pain is to be distributed, whether to raise taxes or cut spending on this or that group, and what the income-distribution effects of their policies are. Those are political questions, not technocratic ones. ("Have PhD, Will Govern," 2011)

The issue of how science and scientists inform the public and contribute to public policy making in the contemporary United States when there is increasing ideological value and distrust in scientific and democratic institutions will continue to be a pivotal issue. And as Schneider argues, any long-term ability to develop "citizen-scientists" capable of participating in and making scientific policy decisions will require a level of scientific literacy that will "empower citizens to begin to pick a scientific signal out of the political noise that all too often paralyzes the policy process" (2000: 119). This may well be very difficult given Pion and Lipsey's conclusion that "the general public holds a rather simple, stereotyped image of scientists as people and, further, does not consistently differentiate science from technology" (1981: 305). We will return to this topic after reviewing the overall findings of the book concerning how ideology and values impact public orientations toward the case study policy issues presented in previous chapters.

IDEOLOGY, VALUES, SCIENCE, AND POLICY ISSUES

A 2013 editorial by Michael Shermer in *Scientific American* comments that "whereas conservatives obsess over the purity and sanctity of sex, the left's sacred values seem fixated on the environment, leading to an almost religious fervor over the purity and sanctity of air, water and especially food." Shermer argues that liberals are more antiscience on issues related to nature (e.g., GMOs, nuclear energy) while conservatives are more antiscience concerning social issues such as reproductive science and stem cells. In general, this is what was found in the West Coast public survey. Table 7.1 provides an overview of the survey findings for each of the issues covered in the book. Liberals were indeed more skeptical of GMOs when compared to conservatives—they were less likely to consume GMOs, more concerned about the safety of GMOs, and less likely to believe GMOs reduce pesticide use. Similarly, conservatives were found to be more skeptical about technology and teen sex activity when compared to liberals. Conservatives were more likely than liberals to

Table 7.1 Political Ideology and Scientific Controversies: Liberals vs. Conservatives

	Liberal	Conservative
Genetically Modified Organisms	*More skeptical*: Less likely to consume GMOs; more concern about GMO safety; do not believe GMOs reduce pesticide use.	*Less skeptical*: More likely to consume GMOs; more likely to believe GMOs are safe; more likely to believe GMOs reduce pesticide use.
Vaccinations	*More skeptical*: More concern about vaccine harmful side-effects; more concern there is risk from vaccines; less likely to feel vaccines safe.	*Less skeptical*: Less concern about vaccine harmful side-effects; less concern there is risk from vaccines; more likely to feel vaccines safe.
Climate Change	*Less skeptical*: More likely to believe Earth getting warmer; more likely to believe warming is human caused.	*More skeptical*: Less likely to believe Earth getting warmer; less likely to believe warming is human caused.
Teen Sex	*Less skeptical*: Less likely to believe HPV vaccination will increase teen sexual activity; less likely to believe Plan B will increase teen sexual activity; less likely to believe teaching abstinence will reduce teen sexual activity.	*More skeptical*: More likely to believe HPV vaccination will increase teen sexual activity; more likely to believe Plan B will increase teen sexual activity; more likely to believe teaching abstinence will reduce teen sexual activity.
Stem Cells	*Less skeptical*: More likely to agree that stem cells are acceptable in medical research; more likely to support use of human embryo stem cells.	*More skeptical*: Less likely to agree that stem cells are acceptable in medical research; less likely to support use of human embryo stem cells.

believe that the HPV vaccination and Plan B would increase teen sex activity, and that teaching abstinence in schools would reduce teen sex activities.

As expected there were also ideological differences for climate change and stem cells. Liberals were more likely to believe that the earth is getting warmer and that humans are responsible when compared to conservatives. Liberals were also less skeptical about stem cells than conservatives, with conservatives less likely to agree that stem cells are acceptable in medical research and less likely to support the use of human ESCs in medical research when compared to liberals. While the literature suggests that ideologues on both the left and right may be opposed to certain immunizations, the public survey data reported in this study found that conservatives when compared to liberals were less concerned about the harmful side of immunizations, less concerned about whether there are risks associated with vaccinations, and more likely to feel that vaccines are safe.

This study also examined the impact of postindustrial values on the case study policy issues. Postindustrial values have been found to be associated with some of the most cynical and penetrating critiques concerning science and the scientific method related to the postmodern perspective. Postmodernism is primarily concerned with the validity of "truth claims." Postmodernists argue that in contemporary societies, knowledge is incorrectly equated with science and the scientific method; other forms of insight such as *narratives* (e.g., oral histories, individual perspectives, ethnographic accounts, etc.) are considered to be inferior to science because they are "subjective" and/or "soft" forms of information often premised on anecdotal evidence. They further argue that virtually all formal language and nearly all prominent abstract ideas are fundamental expressions of power, and that scientific writings are little more than a clever means used to reinforce the "authority of the powerful"—namely, Western culture, and most especially that of privileged white males (Latour, 2005; Marx, 1994). Postmodernists reject positivism and the scientific method, and they argue that it has no special or privileged claim to truth (i.e., the accurate depiction of reality) (Kuntz, 2012). While the philosophy of postmodernism is mostly in the realm of academia, postmaterialist values are the public manifestation of the move to postmodernism, according to Inglehart (1997).

Table 7.2 displays the impact of postmaterialist values on the policy controversies covered in this book. While the results from the West Coast survey were not as pronounced as the ideological differences investigated, there are still some clear patterns in the survey data. When compared to materialists, postmaterialists were found to be more skeptical of GMOs and vaccinations, but less skeptical of climate change, the impact of reproductive technologies, and abstinence on teen sex activities, and less skeptical concerning the use of stem cells in research.

More specifically, postmaterialists were more skeptical of GMOs than materialists and are less likely to consume GMOs, are more concerned about GMO safety, and do not believe that GMOs reduce pesticide use. When compared to materialists, postmaterialists were also more

Table 7.2 Postmaterialist Values and Scientific Controversies: Postmaterialists vs. Materialists

	Postmaterialist	Materialist
Genetically Modified Organisms	*More skeptical*: Less likely to consume GMOs; more concern about GMO safety; do not believe GMOs reduce pesticide use.	*Less skeptical*: More likely to consume GMOs; more likely to believe GMOs are safe; more likely to believe GMOs reduce pesticide use.
Vaccinations	*More skeptical*: More concern about vaccine harmful side-effects; more concern there is risk from vaccines; less support for universal vaccination if others are vaccinated; less likely to feel vaccines safe.	*Less skeptical*: Less concern about vaccine harmful side-effects; less concern there is risk from vaccines; more support for universal vaccination if others are vaccinated; more likely to feel vaccines safe.
Climate Change	*Less skeptical*: More likely to believe Earth getting warmer; more likely to believe warming is human caused.	*More skeptical*: Less likely to believe Earth getting warmer; less likely to believe warming is human caused.
Teen Sex	*Less skeptical*: Less likely to believe HPV vaccination will increase teen sexual activity; less likely to believe Plan B will increase teen sexual activity; less likely to believe teaching abstinence will reduce teen sexual activity.	*More skeptical*: More likely to believe HPV vaccination will increase teen sexual activity; more likely to believe Plan B will increase teen sexual activity; more likely to believe teaching abstinence will reduce teen sexual activity.
Stem Cells	*Less skeptical*: More likely to agree that stem cells are acceptable in medical research; more likely to support use of human embryo stem cells.	*More skeptical*: Less likely to agree that stem cells are acceptable in medical research; less likely to support use of human embryo stem cells.

concerned about the side effects of vaccines, more concerned about vaccine risks, less supportive of universal vaccination, and less likely to feel vaccines are safe. Perhaps postmaterialists' skepticism in comparison to other values types concerning GMOs and vaccinations is based on their suspicion of large commercial agriculture (e.g., Monsanto) and Big Pharma as large multinational corporations. This explanation would be consistent with previous research that found that Canadian and U.S. postmaterialists are significantly less trusting of corporations and other powerful private sector interests than other value types (Steel, Lovrich, and Pierce, 1994).

As for the remaining policy issues, postmaterialists are more consistent with the orientations of ideological liberals, and materialists are more consistent with ideological conservatives. Concerning climate change, postmaterialists when compared to materialists are more likely to believe the earth is getting warmer and that it is the result of human activity. Postmaterialists are also less likely than materialists to believe that HPV vaccination and Plan B will increase teen sex activity and that teaching abstinence will reduce teen sexual activity. Finally, postmaterialists when compared to materialists are more likely to agree that stem cells are acceptable in medical research and more likely to support human embryo–based research.

The third set of values examined for perhaps the strongest and most unquestioning supporters of the ability of science and the scientific method to predict various phenomena in the physical and social world accurately and objectively are adherents to the school of "logical positivism." While logical positivism has evolved into a highly diverse school of thought with multiple perspectives, there are some common themes prevalent among many adherents. They include: (1) science can provide universal truths about our world; (2) the knowledge produced by science can be objective; (3) scientific knowledge is linear and can lead to human progress; and (4), scientists must be free to conduct research in an open society within certain ethical considerations. As Babbie describes it, "Positivism has generally represented the belief in a logically ordered, objective reality that we can come to know" (1998: 50). While few contemporary scientists would completely agree with the tenets of positivism of a logically ordered, objective reality that can be understood through the tireless application of the scientific method, as the West coast survey revealed in Chapter 1a large portion of the public does agree, and this perception of science does have a significant impact on how the public views the various policy issues investigated in this book.

When examining positivist beliefs about science and the impact on orientations toward GMO, vaccinations, climate change, teen sex, and stem cells, the West Coast survey results found that a high level of belief in positivism is consistent with scientific consensus on the issues (see Table 7.3). When compared to low-level believers in positivism, high-level believers were more supportive of GMOs; more supportive of vaccinations; more likely to believe the earth is warming and that humans are responsible; less likely to believe Plan B and HPV vaccination lead to

Table 7.3 Positivism Beliefs and Scientific Controversies: Low vs. High Levels of Positivism Beliefs

	Low	High
Genetically Modified Organisms	*More skeptical*: Less likely to consume GMOs; more concern about GMO safety; do not believe GMOs reduce pesticide use.	*Less skeptical*: More likely to consume GMOs; more likely to believe GMOs are safe; more likely to believe GMOs reduce pesticide use.
Vaccinations	*More skeptical*: More likely to believe vaccines cause autism; more likely to believe vaccines have harmful side-effects; more concern there is risk from vaccines; less support for universal vaccination if others are vaccinated; less likely to feel vaccines safe.	*Less skeptical*: Less likely to believe vaccines cause autism; less likely to believe vaccines have harmful side-effects; less concern there is risk from vaccines; more support for universal vaccination if others are vaccinated; more likely to feel vaccines safe.
Climate Change	*More skeptical*: Less likely to believe Earth getting warmer; less likely to believe warming is human caused.	*Less skeptical*: More likely to believe Earth getting warmer; more likely to believe warming is human caused.
Teen Sex	*More skeptical*: More likely to believe HPV vaccination will increase teen sexual activity; more likely to believe Plan B will increase teen sexual activity; more likely to believe teaching abstinence will reduce teen sexual activity.	*Less skeptical*: Less likely to believe HPV vaccination will increase teen sexual activity; less likely to believe Plan B will increase teen sexual activity; less likely to believe teaching abstinence will reduce teen sexual activity.
Stem Cells	*More skeptical*: Less likely to agree that stem cells are acceptable in medical research; less likely to support use of human embryo stem cells.	*Less skeptical*: More likely to agree that stem cells are acceptable in medical research; more likely to support use of human embryo stem cells.

increased teen sexual activity; less likely to believe teaching abstinence will reduce teen sexual activity; and more likely to believe stem cells are acceptable in medical research and support the use of human embryos in research.

Given how values and ideology shape public perceptions of science and scientific issues, what should be the role of science and scientists in the policy process and how can and should the public consider science as a reputable source to policy? These are the questions that will be covered in the next section.

SCIENCE, THE PUBLIC, AND THE POLICY PROCESS

According to Underdal, the ability of scientists to influence the policy process is dependent on two key factors, assumptions about their integrity and perceptions of their competence (2000: 10). Underdal further argues that the more autonomy scientists have to conduct their research, the more integrity their results will have with decision makers. If scientists are funded to produce specific results by government or business, the integrity and thus utility of their results will be diminished (see Table 7.4). In addition, scientists that are recruited because of their scholarly merits and publication records will have more credibility, and the research results produced will have more integrity if scientists are allowed to pursue their own research agendas.

How likely is it that scientists and science will be utilized in the policy-making process? Underdal has suggested that a number of factors likely determine the degree to which science will be used in the policy process. A partial list of these factors is set forth in Table 7.5. Science is most likely to be used in the policy-making process if there is a feasible or practical cure available for the problem; if there is a public and scientific consensus on the definition and description of the problem; if the effects of the problem are immediate in time; if the problem affects the social and economic center of society; if the policy problem is developing rapidly; if the effects of the problem are apparent or visible to the public; and finally, if there is little political conflict over the resolution of the issue.

Table 7.4 Scientific Autonomy and Integrity

High Autonomy and Integrity of Science	*Low Autonomy and Integrity of Science*
Research funded by scientific organization.	Research funded by business, government, or party interested in application of results.
Scientists recruited on scholarly merits or role in scientific community.	Scientists recruited on basis of political orientations.
High level of research autonomy; scientists set own research agenda and organize own work.	Under effective control of business, agency of government, or party interested in application of results.

Source: Revised from Underdal, 2000: 13.

Table 7.5 Conditions Affecting the Impact of Scientific Information on Public Policy

Impact of Science on Policy Likely to Be Strong	*Impact of Science on Policy Likely to Be Weak*
Consensus on definition and description of problem.	Tentative or contested descriptions of problem.
Feasible or practical "cure" available.	"Cure" unclear or not feasible.
Effects close in time.	Effects remote.
Problem affecting social and economic center of society.	Problem affecting periphery.
Problem developing rapidly and surprisingly.	Problem developing slowly and according to expectations.
Effects experienced by, or at least visible to, the public.	Effects not yet experienced by, or visible to, the public.
Political conflict low.	Political conflict high.

Source: Revised from Underdal, 2000: 16.

Of course, while there is a consensus among scientists on most if not all of the issues discussed in this volume, there is much political conflict and less public consensus depending on the issue examined due to ideological and value perspectives. The policy issues examined here mostly reflect Underdal's second category of characteristics of a weak impact of science on policy in many contexts given the level of political conflict and potentially the lack of an immediate or visible threat for many issues. How then can science and scientists be incorporated into the policy process given this high degree of ideological polarization and value polarization?

Often many proponents of science-based policy talk about "educating the public" and increasing the scientific literacy of the public. However, as Douglas cautions, "more knowledge of a topic does not automatically lead to more acceptance of the science or its implications" (2015: 297). In her review of research conducted by Bolsen, Druckman, and Cook (2015), Douglas further commented that "deeper knowledge among members of the public about climate change science did not translate into deeper acceptance of the scientific consensus. Instead, it depended on the person's ideological commitments" (2015: 297).

The work of Kuklinski et al. (2000) reminds us that there is a fundamental difference between those among the public who are uninformed and those who are misinformed. They argue that it is possible to change public positions on issues by providing relevant information and facts, but that many citizens who are misinformed base their positions on misleading and inaccurate information they believe is true and consistent with their political preferences. In other words, due to shortcuts of motivated reasoning, misinformation is easily acquired if from a source that adheres to preheld beliefs.

Why is it necessary to engage in motivated reasoning? Essentially, "on their own, individuals are not well equipped to separate fact from fiction, and they never will be. Ignorance is our natural state; it is a product of the way the mind works" (Fernbach and Sloman, 2017). Human physical evolution ran concurrent with cultural evolution. Therefore the ability to gain knowledge from a collective group of people allowed humans to specialize tasks instead of requiring laborious decision making as an autonomous individual. Thus humans rely on rules, norms, and, perhaps most relevant, shared information to guide societal structure. Fernbach and Sloman (2017) suggest that it is our shared information that makes humans unique:

> What really sets human beings apart is not our individual mental capacity. The secret is our ability to jointly pursue complex goals by dividing cognitive labor. Hunting, trade, agriculture, manufacturing—all of our world-altering innovations—were made possible by this ability.

It is this shared knowledge though that impacts our policy choices, and it is not necessarily premised on scientific facts or knowledge.

The consequences of cognitive biases and motivated reasoning are that policy decisions can be based on nonsensical thinking, not rational or evidence-based decision making. Further, this amassed knowledge perpetuates policy decisions that can have severe consequences for humans, specifically when considering issues like climate change. Of this phenomenon, Fernbach and Sloman (2017) state:

> Such collective delusions illustrate both the power and the deep flaw of human thinking. It is remarkable that large groups of people can coalesce around a common belief when few of them individually possess the requisite knowledge to support it . . . the same underlying forces explain why we can come to believe outrageous things, which can lead to equally consequential but disastrous outcomes.

While some researchers argue that uninformed citizens can use heuristics as a substitute for factual information from experts (Sniderman, Brody, and Tetlock, 1991; Lupia and McCubbins, 1998), other researchers suggest a more problematic dynamic concerning the ability to overcome ideology and values in this regard. The research by Nyhan and Reifler (2010), for example, casts doubt on the ability to correct citizen ignorance and/or misunderstanding of science and scientific findings. Using experimental design to investigate the ability to correct "unsubstantiated beliefs" among the public concerning various policy issues, Nyhan and Reifler conclude that "corrections frequently fail to reduce misperceptions among the targeted ideological group. We also document several instances of a 'backfire effect' in which corrections actually *increase* misperceptions about the groups in question" (emphasis in original; 2010: 303). The results of these findings led the authors to conclude that citizens engage in motivated reasoning when forming policy references, and that it will be most difficult to provide corrective information that would lead to science-based policy preferences.

Various approaches have been suggested to overcome the problem of ideological and value-driven motivated reasoning concerning important policy issues when there is scientific consensus on the issue. Nisbet and Mooney in a *Science* commentary argue (2007): "Without misrepresenting scientific information on highly contested issues, scientists must learn to actively 'frame' information to make it relevant to different audiences." The authors explain that citizens are literally bombarded with so much information from so many different sources that they screen and select information sources that are more consistent with their own ideological and value dispositions. Therefore to have them consider scientific information concerning contentious issues, the information provided should be framed in a manner that resonates with those ideological and value systems. An example of such framing could be similar to how renewable energy technologies have been promoted in the American west, where wind farms have been lauded as non-fossil-fuel-based technologies that will "decrease green-house gases" among urban and more liberal populations who are concerned about climate change, and as an "economic development" opportunity among rural and more conservative populations who are not concerned about and/or do not believe in climate change (Pierce and Steel, 2017).

Nisbet and Mooney raise the issue that the framing of science may seem manipulative and "Orwellian" (or, perhaps, Lysenkoist) in nature; however, they justify the approach on the basis that "facts will be repeatedly misapplied and twisted in direct proportion to their relevance to the political debate and decision-making" (2007). They also argue that scientists should avoid overly technical details of scientific research when trying to promote and defend it. Careful framing and presentation of information is integral to having an impact on relevant ideological groups.

Often, social science focuses on how to correct misinformation on the individual level; thus framing of issues is targeted toward individuals. Deva R. Woodly suggests that enacting change on a policy may require appealing to the public generally, not individuals, since motivated reasoning implies collective understanding. Woodly asserts that "public opinion research on priming and framing has shown that even though people do not easily change their personal attitudes, differences in the topics and frames discussed as part of the regular public debate do change people's perceptions of issue salience as well as what is at stake in political debates" (2015: 10). By examining two policy movements, marriage equality and the living wage, Woodly finds that marriage equality advocates found success by altering the framing away from the act of homosexuality and "traditional" marriage "toward an emphasis on the fundamental similarity of individuals who form romantic and familial bonds according to their own 'orientation' and their attendant civil rights" (2015: 11). As the public discourse around marriage equality took on these talking points, acceptance became more widespread. Conversely, arguments for the living wage "remained variable and inconsistent" (Woodly, 2015: 11) and therefore advocates remained stuck in a quagmire of defending their position rather than gaining policy traction through an evolved "resonant" frame that presented the issue consistently and coherently (Woodly, 2015: 11).

Framing of issues may also resonate more if initiated by groups that share ideological or other worldviews. Recently, some Republican leaders in Congress joined the Climate Solutions Caucus, a group of both Democrats and Republicans working on bipartisan efforts to combat climate change that "is bolstered by a new chorus of big business, faith groups and young college-based Republicans that are demanding that the GOP drops the climate skepticism that has become a key part of its identity over the past decade" (Milman, 2017). This Republican leadership, in conjunction with conservative groups like the Partnership for Responsible Growth who are targeting climate change campaigns to engage Republicans on climate action, may be more successful in conveying the urgency of climate change and encouraging policy support from a more conservative base.

Another approach suggested by Douglas is that there should be an effort to "educate the public on the nature of science" and not focus more narrowly on the science on specific contentious issues (2015: 305). She argues that this might best happen through the use of "better social forums" for discussing and debating scientific issues that include both scientists and the public. These forums should focus on questions and dialogue such as: "Why do you think like you think? How do you respond to the evidence that goes against what you think?" (2015: 305). This approach goes beyond the "normal science" approach to science and the policy process discussed in Chapter 1, where scientists play a minimalist role in the policy process by just conducting and publishing their research and then leaving it to others to use or not use their research findings.

This traditional approach to science has come under increasing scrutiny in recent years as an optimal model for science-based policy making. Some observers have called this the "science wars" in the United States where experts have been pitted against local knowledge and practices (see Ross 1996). The traditional model of science is seen by many as inadequate for various reasons. The combination of the complexity of the problems faced and the known limits of human measurement and analytical abilities come together to constrain the power of science in this area of governmental responsibility. As Funtowicz and Ravetz (1999) noted from their own work concerning environmental policy:

> Anyone trying to comprehend the problems of "the environment" might well be bewildered by their number, variety and complication. There is a natural temptation to try to reduce them to simpler, more manageable elements, as with mathematical models and computer simulations. This, after all, has been the successful programme of Western science and technology up to now. But environmental problems have features that prevent reductionist approaches from having any but the most limited useful effect. These (features) are what we mean when we use the term "complexity." Complexity is a property of certain sorts of systems; it distinguishes them from those that are simple, or merely complicated. Simple systems can be captured (in theory or in practice) by a deterministic, linear causal analysis.

A second, emerging model challenges the traditional and normal role of science and scientists, not so much on the authority of scientific information and the acceptability of positivism but on the proper roles for research scientists in policy making and management (Kay, 1998). It proposes that scientists should become more integrated into management and policy processes. Research scientists need to come out of their labs and in from their field studies to directly engage in public environmental decisions within natural resource agencies and such venues as courts and public hearings. There is a need for more science in these processes and decisions, this approach argues, but this can be brought about only if scientists themselves become more actively involved. Moreover, this model suggests that scientists should not hesitate to make judgments about policy alternatives, if the preponderance of evidence and their judgment moves them in certain practical directions. They are, after all, in the best position to interpret the scientific data and findings and thus are in a special position to advocate for specific policies and alternatives.

This emerging "postnormal science" approach calls for personal involvement by individual scientists in public policy decision making, providing expertise and even promoting specific strategies that they believe are supported by the available scientific knowledge (Steel and Weber, 2001). Funtowicz and Ravetz (1999) articulated this model as follows:

> There is a new role for natural science. The facts that are taught from textbooks in institutions are still necessary, but are no longer sufficient. For these relate to a standardized version of the natural world, frequently to the artificially pure and stable conditions of a laboratory experiment. The world as we interact with it in working for sustainability is quite different. Those who have become accredited experts through a course of academic study, have much valuable knowledge in relation to these practical problems. But they may also need to recover from the mindset they might absorb unconsciously from their instruction. Contrary to the impression conveyed by textbooks, most problems in practice have more than one plausible answer; and many have no answer at all.

An example of the postnormal approach for science and scientists is Stankey and Shindler's (2006) work with management policies for "rare and little-known species" (RLKS) such as snails, fungi, and slugs found in the forests of the U.S. Pacific Northwest. The authors found that limited public awareness and knowledge of the science of RLKS led to public resistance to science-based policies and therefore a decline in species. This study found that while there had been much scientific research on the economic and biological aspects of RLKS policy, there had been no research whatsoever to "foster cultural adoptability, or as commonly termed, social acceptability" of management plans among the public (2006: 29). In order to enhance public understanding and participation in RLKS management, Stankey and Shindler took a proactive and postnormal approach by having scientists, managers, and community members engage in joint fact finding and collaborative discovery. They argue that the key to social acceptability and participation for successful management requires collaborative approaches that: (1) clarify the rationale and potential impacts of policies on species and communities; (2) outline specific actions that will be taken with the management plan; (3) specify and adapt to the contextual setting of the issue; and (4) identify where and when management plans will be implemented (2006: 28).

However, a limitation of this approach of a more "integrative role" for scientists is that they operate in a communal scientific environment that imposes different demands on their time and energy, and their reputations and identities as scientists depend upon a different system of institutional relationships and rewards (e.g., academic scientists are rewarded for publications and grant writing and not necessarily public outreach and engagement). Involvement in the public policy process requires somewhat different communication and interpersonal skills than those that are effective in the scientific community. It may also elicit normative opinions in the scientific and policy arenas that can undermine scientists' authority and personal decorum (i.e., engagement in policy advocacy). Other scientists sometimes have reservations about researchers who do become involved in policy matters, and may question their standing and credibility as a result. These and other factors can mean that scientists will be wary of researchers taking a more active, integrative role in policy making. As Jamieson argued, "What most scientists want to do is (relatively) basic science: they want to discover the most fundamental particle, understand the human genome, the atmospheric system, the immune system, and so on" (2000: 322).

Given the difficulties of outreach efforts because of ideological and value orientations, and the logistics and limited ability to reach wider public audiences using more localized postnormal science approaches, another option is to focus on ideological moderates among the public to engage with. The West Coast survey results indicate that moderates—including moderate liberals and moderate conservatives—are more receptive of science and scientific information in comparison to ideological liberals and conservatives. While the largest self-identified ideological group in the United States are conservatives, if you combine moderates, moderate conservatives, and moderate liberals together, they outnumber ideological conservatives and ideological liberals in the U.S. context (Gallup, 2016). It may make sense for scientists to focus on that segment of the citizenry to engage with to promote scientific research and literacy, as they tend to see the complexities of issues from both a liberal and conservative angle (Ball, 2014). Of course, as this study has also shown, not all liberals and conservatives take antiscience positions on all issues as well, so perhaps building coalitions of moderates with ideological

Table 7.6 Political Ideology and Belief in Positivism

Ideology	Liberal	Moderate	Conservative
Belief in Positivism			
Low	36.7%	31.2%	39.1%
Medium	26.8%	40.2%	36.5%
High	36.4%	28.6%	24.4%
N =	686	381	394

Chi-square = 31.724, $p = .000$

conservatives and liberals who are more open to scientific research may be a productive approach.

The West Coast survey data presented in Tables 7.6 and 7.7 show the relationship between political ideology and belief in positivism. Table 7.6 shows that liberals and moderates are slightly yet significantly more likely to have high levels of belief in positivism when compared to conservatives. Over 36 percent of liberals and 28.6 percent of moderates have high levels of belief in positivism, compared to 24.4 percent of conservatives. The data in this table grouped all respondents who called themselves "liberal" and "conservative" in the survey regardless of whether they were moderate or more ideological on the scale that was used in the survey (1 = very liberal to 9 = very conservative). However, the data displayed in Table 7.7 include the percentage of respondents who said they are ideologically "moderate" or moderately liberal or moderately conservative who also have high levels of belief in positivism (including those who responded 5 = moderate, 4 = moderately liberal, and 6 = moderately conservative). Over 67 percent of respondents fit this classification, which by far outnumbers ideological respondents with lower levels of belief in positivism, providing a potentially large and receptive public audience for science in the West Coast context.

Table 7.7 Ideology and Belief in Positivism Combined

	Percent	Frequency
Ideological and Lower Levels of Belief in Positivism	31.6%	470
Ideologically Moderate and High-Level Belief in Positivism	67.2%	998

SUMMARY

We live in a society exquisitely dependent on science and technology, in which hardly anyone knows anything about science and technology.

—Carl Sagan

While this chapter provided a few ideas about how to engage the public on scientific research given ideological and value polarization, there are most likely many more ideas worth pursuing. However, as a former regional director of the National Marine Fisheries Service commented: "Most people practice pick-and-choose . . . agenda-driven science in which the quality of the science is judged by the apparent results achieved. This is not biological science but political science" (cited in Blumm, 2002: 327). As we began this book, the new Trump administration has clearly launched an almost unprecedented attack on science and scientists and has appointed a mostly antiscience cabinet (*New York Times* Editorial Board, 2017). And, as we found in the analyses presented here, ideology and values are important considerations in how the public *and* political elites process scientific information and form opinions on various policy issues.

The implication of these findings for the use of science in the policy process suggests continuing ideological and value polarization. While moderates, some conservatives, and liberals were supportive of science and scientists, many on the right and some on the left remain skeptical of scientific findings depending on the issue. One factor that was not investigated in this study, and for which data were not collected in the survey, is that of religious orientations. There is much research showing that religion and especially fundamentalist religious values have an impact on orientations toward issues like teen sex and stem cells, but these views are also strongly associated with conservative political orientations, which was investigated in this study (Tatalovich and Daynes, 2011).

While scientists are expressing shock at the most recent examples of political manipulation of science by the new Trump administration, social scientists have been observing the politicization of science for decades (e.g., Mooney, 2005; Beck, 1992; Dickson, 1984; Primack and von Hippel, 1974). Some scientists, for example, accused President Nixon of inaccurately portraying that his scientific advisors supported a research program concerning a proposed supersonic transport airplane when in fact many had serious concerns that the plane would damage the ozone layer and contribute to climate change (Dickson, 1984). Another example is President Lyndon Johnson's secretary of defense Robert McNamara, who misrepresented scientific information about antiballistic missiles. This was a situation "that left many scientists disenchanted with the role they were being placed in as advisers to the executive branch" (Dickson, 1984: 225).

Others, however, note a tendency for institutional science and individual scientists to adapt to changing political whims and ideological priorities, securing funding by reframing and reconfiguring research questions, topics, and methods (Collingridge and Reeve, 1986). For example, Beck (1992) noted that one of the trade-offs for this "politicization of science" was the "scientization of policy." In the latter half of the twentieth century, institutional science in the developed world has itself been involved in the control of research agendas and funding strategies for programs. Findings that emerge from these research agendas have been used to rationalize policy and practices based on the value and ideological priorities of the funders (Greenberg, 2001), thereby further confounding the ability of science and scientists to contribute to the debate.

In the early 1970s, there was an attempt to promote public engagement and literacy in science by liberal groups through the use of "public interest science" (Primack and von Hippel, 1974). Although Congress authorized and appropriated funds for a "Science for Citizens" program in 1976, there was little enthusiasm for the idea at the NSF or among many in the mainstream scientific community who "objected to its activist tone" (Dickson, 1984: 230). Therefore the program faded away quietly by the end of the decade. However, as discussed above, there has been a resurgence of interest in "public interest science" through postnormal approaches, and also a variant by conservative groups who are increasingly convinced that their values about the impact of scientific advances like cloning and genetic modification and about evolutionary theory are being left out of the policy decisions. Perhaps further direct engagement by the pubic in citizen science and postnormal approaches in the twenty-first century will lead to increased scientific literacy and enhanced consideration of scientific information by the public. If the March for Science held on Earth Day (April 22) 2017—with participation on six continents and thousands marching in cities throughout the United States and Western Europe in protest of the new administration's proposed cuts to scientific research and its climate change denial perspective —is any indication, the public may well be poised and activated to engage in scientific inquiry and science-based policy. However, only time will tell, and as discussed throughout the book, it will be difficult to get through the "rose-tinted lenses" of ideology and values.

In conclusion, we believe that scientists need to work with and engage all groups and citizens—and particularly moderates and those that lean moderate—to build the overall level of trust and literacy in both science and scientists. While almost all scientists and most nonscientists believe that scientists should not make policy, there is an emerging preference for an "integrative" approach to science. This "postnormal science" calls for personal involvement by individual research scientists in bureaucratic and public decision making, providing expertise and promoting specific strategies that they believe are supported by the available scientific knowledge (Ravetz, 1987; Steel and Weber, 2001). Others, such as Kai Lee, have similarly called for a new "civic science" to integrate science and scientists in the policy process (Lee, 1993). While these approaches may put scientists into very uncomfortable situations outside of the laboratory and into the political realm, they will serve to familiarize both nonscientists with the strengths and limitations of science in policy making and scientists with the "sausage making" of policy.

NOTE

1. M. L. Finkel, *Truth, Lies and Public Health: How We Are Affected When Science and Politics Collide* (Westport, CT: Praeger, 2007), 12.

REFERENCES

Babbie, E. *The Practice of Social Research*. 8th ed. Belmont, CA: Wadsworth, 1998.

Ball, M. "Moderates: Who Are They and What Do They Want?" *Atlantic*, May 15, 2014.

Beck, U. *Risk Society: Towards a New Modernity*. London: Sage, 1992.

Blumm, M. *Sacrificing Salmon: A Legal and Policy History of the Decline of Columbia River Salmon*. Portland, OR: BookWorld Publications, 2002.

Bolsen, S., J. N. Druckman, and F. L. Cook. "Citizens', Scientists', and Policy Advisors' Beliefs about Global Warming." *ANNALS of the American Academy of Political and Social Science* 658 (2015): 271–95.

Collingridge, D., and C. Reeve. *Science Speaks to Power: The Role of Experts in Policy Making*. New York: St. Martin's Press, 1986.

Dickson, D. *The New Politics of Science*. New York: Pantheon Books, 1984.

Douglas, H. "Politics and Science: Untangling Values, Ideologies, and Reasons." *ANNALS of the American Academy of Political and Social Science* 658 (2015): 296–306.

Fernbach, P., and S. Sloman. "Why We Believe Obvious Untruths." *New York Times*, March 3, 2017.

Finkel, M. L. *Truth, Lies, and Public Health: How We Are Affected When Science and Politics Collide*. Westport, CT: Praeger, 2007.

Funtowicz, S., and J. Ravetz. "Post-Normal Science: Environmental Policy under Conditions of Complexity." NUSAP.net (1999).

Gallup. "Conservatives Hang on to Ideology Lead by a Thread." 2016.

Greenberg, D. S. *Science, Money, and Politics: Political Triumph and Ethical Erosion*. Chicago: University of Chicago Press, 2001.

"Have PhD, Will Govern." *Technocrats and Democracy* (blog), *Economist*, November 16, 2011.

Inglehart, R. *Modernization and Postmodernization: Cultural, Economic, and Political Change in 43 Societies*. Princeton, NJ: Princeton University Press, 1997.

Jamieson, D. "Prediction in Society." In *Prediction: Science, Decision Making, and the Future of Nature*, edited by D. Sarewitz, R. A. Pielke, and R. Byerly, 315–26. Washington, DC: Island Press, 2000.

Kuklinski, J. H., P. J. Quirk, J. Jerit, D. Schweider, and R. F. Rich. "Misinformation and the Currency of Democratic Citizenship." *Journal of Politics* 62 (2000): 790–816.

Kuntz, M. "The Postmodern Assault on Science." *European Molecular Biology Organization* 13 (2012): 885–89.

Latour, B. *Science in Action: How to Follow Scientists and Engineers through Society*. Cambridge, MA: Harvard University Press, 2005.

Lee, K. *Compass and Gyroscope*. Washington, DC: Island Press, 1993.

Levien, R. "Global Problems: The Role of International Science and Technology Organizations." In *Science, Technology and Global Problems*, edited by J. Gvishiani, 45–50. Oxford: Pergamon Press, 1979.

Lupia, A., and M. McCubbins. *The Democratic Dilemma: Can Citizens Learn What They Need to Know?* New York: Cambridge University Press, 1998.

Lynas, M. "Even in 2015, the Public Doesn't Trust Scientists." *Washington Post*, January 30, 2015.

Marx, L. "The Environment and the 'Two Cultures' Divide." In *Science, Technology, and the Environment: Multidisciplinary Perspectives*, edited by J. R. Fleming and H. Gemery, 3–21. Akron, OH: University of Akron Press, 1994.

Milman, O. "The Republicans Who Care about Climate Change: 'They Are Done with the Denial'." *Guardian*, April 27, 2017.

Mooney, C. *The Republican War on Science*. Cambridge, MA: Basic Books, 2005.

National Research Council. *Using Science as Evidence in Public Policy*. Washington, DC: National Academies Press, 2012.

New York Times Editorial Board. "The Trump Administration's War on Science." *New York Times*, March 27, 2017.

Nisbet, M. C., and C. Mooney. "Framing Science." *Science* 316 (2007).

Nyhan, B., and J. Reifler. "When Corrections Fail: The Persistence of Political Misperceptions." *Political Behavior* 32 (2010): 303–30.

Pew Research Center. "Beyond Distrust: How Americans View Their Government, 1958–2015." 2015.

Pierce, J. C., and N. P. Lovrich. "The Technocracy versus Democracy Quandary." In *Science and Politics: An A-to-Z Guide to Issues and Controversies*, edited by B. S. Steel, 134–38. Thousand Oaks, CA: Sage/CQ Press, 2014.

Pierce, J. C., and B. S. Steel. *Prospects for Alternative Energy Development in the U.S. West: Tilting at Windmills?* Dordrecht, Netherlands: Springer Press, 2017.

Pion, G., and M. Lipsey. "Public Attitudes toward Science and Technology: What Have the Surveys Told Us?" *Public Opinion Quarterly* 45 (1981): 303–16.

Primack, J., and F. von Hippel. *Advise and Dissent: Scientists in the Political Arena.* New York: Basic Books, 1974.

Ravetz, J. *The Merger of Knowledge with Power: Essays in Critical Science.* London: Mansell, 1990.

Ross, A. *Science Wars.* Durham, NC: Duke University Press, 1996.

Schneider, S. "Is the 'Citizen Scientist' an Oxymoron?" In *Science, Technology, and Democracy*, edited by D. L. Kleinman, 103–20. Albany, NY: SUNY Press, 2000.

Shermer, M. "The Liberals' War on Science: How Politics Distorts Science on Both Ends of the Spectrum." *Scientific American*, February 1, 2013.

Sniderman, P. M., R. A. Brody, and P. E. Tetlock. *Reasoning and Choice: Explorations in Political Psychology.* New York: Cambridge University Press, 1991.

Stankey, G., and B. Shindler. "Formation of Social Acceptability Judgments and Their Implications for Management of Rare and Little-Known Species." *Conservation Biology* 20 (2006): 28–37.

Steel, B. S., P. List, D. Lach, and B. Shindler. "The Role of Scientists in the Environmental Policy Process: A Case Study from the American West." *Environmental Science and Policy* 7 (2004): 1–13.

Steel, B. S., N. P. Lovrich, and J. C. Pierce. "Trust in Natural Resource Information Sources and Postmaterialist Values: A Comparative Analysis of U.S. and Canadian Citizens in the Great Lakes Area." *Journal of Environmental Systems* 22 (1994): 123–36.

Steel, B. S., and E. Weber. "Ecosystem Management, Devolution, and Public Opinion." *Global Environmental Change* 11 (2001): 119–31.

Tatalovich, R., and B. Daynes, eds. *Social Regulatory Policy.* New York: M. E. Sharpe, 2011.

Underdal, A. "Science and Politics: The Anatomy of an Uneasy Partnership." In *Science and Politics in International Environmental Regimes*, edited by G. Steinar, A. Skodyin, A. Underdal, and P. Wettestad, 1–21. Manchester, England: Manchester University Press, 2000.

Vigoda, E. "From Responsiveness to Collaboration: Governance, Citizens, and the Next Generation of Public Administration." *Public Administration Review* 62 (2002): 527–40.

Woodly, D. R. *The Politics of Common Sense: How Social Movements Use Public Discourse to Change Politics and Win Acceptance.* New York: Oxford University Press, 2015.

Index

Abstinence-only education, 1; HPV vaccination and, 120; teen sex, 116–19
Accuracy goals, 4–5
Adolescent Family Life Act, 119
Agricultural biotechnology, 30
American Association for Public Opinion Research, 13
American Association for the Advancement of Science, 3
Anthropogenic climate change, 86, 91–92
Antibiotic gene transfers, 31
Antibiotic resistance, 31
Antinuclear movement, 2
Arrhenius, Svante, 86
Asilomar Conference, 30
Association of American Physicians and Surgeons, 62
Atlantic, 59
Attitudes toward science and scientists, 13–16
Autism, 57–58, 61

Babbie, E., 160
Bachmann, Michele, 63, 120
Beliefs, and genetically modified organisms (GMOs), 35–38
Berg, Paul, 30
Bible Belt, 118
Biotechnology, 30
Boudry, M., 7
Braman, D., 5–6
Brown, Jerry, 53–54
Bt *(Bacillus thuringiensis)* corn, 31
Bt *(Bacillus thuringiensis)* cotton, 31
Bush, George W., 138–39, 140

Carson, Rachel, 65
Centers for Disease Control and Prevention (CDC), 122
Cervarix, 112
Chernobyl disaster, 2
Childhood vaccinations, 53–80; analyses, 66–76; controversy, 56–60; ideology, positivism, and postmaterialism, 62–66; overview, 53–56; vaccine skeptics, 60–62
Church of England, 109
Citigroup, 89
Clean Air Act, 90
Climate change, 85–105; analyses, 98–103; anthropogenic, 86, 91–92; defined, 86; direct impacts of, 87–90; human-induced, 93; ideology, positivism, and postmaterialism, 93–98; overview, 85–86;

policies, 94; science and impacts, 86–87; and the U.S. public, 90–92
Climate mitigation *vs.* economy, 92
Climate Solutions Caucus, 166
Clinton, Bill, 55
Cold cognition, 5
Comprehensive sex education, 111, 114
Confirmation bias, 4
Congruence bias, 4
Conservatism, 9
Contraception, emergency, 114–16
Cruz, Ted, 94

Democracy-*vs.*-technocracy quandary, 154–56
Department of Agriculture, 33
Department of Health and Human Services, 139
Dickey-Wicker amendment, 139
Dillman's Tailored Design Method, 13
Directional goals, 5
Disconfirmation bias, 4
Dolly the sheep, 138
Douglas, H., 163, 166
Dunlap, R. E., 97

Economist, 155
Economy *vs.* climate mitigation, 92
Education, abstinence-only, 116–19
Egyptian Empire, 54
Embryonic stem cells: history of, 138–41; politics of, 138–41
Emergency contraception, 114–16
Endangered Species Act, 90
Environmental Protection Agency (EPA), 33
Ethylmercury, 57
European Commission, 32
European Union (EU), 31

Family Research Council, 113
Fernbach, P., 164
FlavrSavr tomato, 31
Focus on the Family, 113, 119
Food allergies, 31
Food and Chemical Toxicity, 33
Food and Drug Administration, 32, 115
Fossil fuel industry, 92
Fox, Michael J., 140
Funtowicz, S., 167–68

Gardasil, 112
Gauchat, G., 8–9
General Mills, 30
General Social Survey, 8
Generation Rescue, 61
Genetically modified organisms (GMOs), 29–49; analyses, 38–45; described, 30–35; discussion, 46–48; ideology, values, and beliefs, 35–38; overview, 29–30
Giddens, Anthony, 104
Gignac, G. E., 63
Global warming, 86
Golden Rice, 34
Grabenstein, J. D., 56
Greenhouse gas (GHG) emissions, 2–3
Greenpeace, 34
Green Revolution, 29
Guardian, 120

Hallman, W.K., 48
Hansen, James, 90
Harvard's Kennedy School of Government, 111, 119
Heritage Foundation, 119
Hiatt, F., 47
Hobby Lobby, 120
Hot cognition, 5
H.R. 810, 139. *See also* Stem Cell Research Enhancement Act of 2005
Human-induced climate change, 93
Human papillomavirus (HPV), 111–14

Ideology, 156–62; childhood vaccinations, 62–66; climate change, 93–98; genetically modified organisms (GMOs), 35–38; stem cells and, 141–43; teen sex, 119–23; and value orientations, 16–23
Industrial Revolution, 87
Inglehart, Ronald, 11, 17, 97
Inglis, Bob, 94
Intergovernmental Panel on Climate Change, 85

Jamieson, D., 169
Jenkins-Smith, H., 6
Johnson, Lyndon, 171
Jones, Abbey M., 60

Kahan, D. M., 6
Kaiser Family Foundation, 111, 119
Kanner, Leo, 57
Kennedy, Robert F., Jr., 9, 59, 63
Krugman, Paul, 94
Kyoto Protocol, 86

Lang, J.T., 48
Levien, R., 154
Lewandowsky, S., 63
Logical positivism, 153, 160
Lynas, Mark, 155

Mad cow disease, 31
Mathematica Policy Research Inc., 121
McCarthy, Jenny, 61
McNamara, Robert, 171
Merck, 112
Millennials, 110
Monsanto, 10, 31–33, 37, 46
Mooney, Chris, 8, 165
Motivated reasoning, 4–7

National Abstinence Education Association, 119
National Cancer Institute, 122
National Environmental Policy Act, 90
NationalMarine Fisheries Service, 171
National Public Radio, 111, 119
National Research Council, 153
National Science Foundation (NSF), 3, 6
National Survey of Family Growth, 110
National Vaccine Compensation Injury Program, 59
Neoliberalism, 94
New York Times, 149
Niles, Meredith, 33
Nisbet, M., 93, 165
Nixon, Richard, 93, 171
Nyhan, B., 165

Obama, Barack, 93, 115, 118, 140
Oberauer, K., 63
Ocean acidification, 87–88
Outcrossing, 31–32

Pap tests, 112
Paris Agreement, 86, 93
Parkinson's disease, 140
Partnership for Responsible Growth, 166
Perry, Rick, 120
Peterson, L. P., 12
Pew Research Center, 3, 9–10, 36, 60, 63, 66, 91–92, 98, 109, 141
Pierce, John, 11
Pigliucci, M., 7
Plan B, 114–16
Policies: issues, 156–62; process, 162–70; *vs.* science, 2–4
Political and value orientations, 7–12
Popper, Karl, 18
Positivism: childhood vaccinations, 62–66; climate change, 93–98; teen sex, 119–23

Postindustrialism, 11
Postmaterialism: childhood vaccinations, 62–66; climate change, 93–98; teen sex, 119–23
Price, Tom, 12, 62–63
Promoter genes, 30
Pruitt, Scott, 105
Public, 162–70
Public health, 32
Puritans, 109

Rabb, Harriet, 139
Ravetz, J., 167
Reagan, Nancy, 140
Reagan, Ronald, 90, 140
Reasoning, motivated, 4–7
Recombinant DNA (rDNA), 30
Reeve, Christopher, 140
Refrigerator mother, 57
Reifler, J., 165
The Republican Brain (Mooney), 8
Rimland, Bernard, 57
Rosi-Marshall, Emma, 33
Roundup Ready canola seeds, 32
RU-486 pill, 114

Schneider, S., 156
Science, 156–62, 162–70; attitudes toward, 13–16; *vs*. policy, 2–4
Scientific American, 156
Scientists, attitudes toward, 13–16
Senate Bill 277, 53
Sexually transmitted diseases (STDs), 1
Shermer, Michael, 156
Shiva, Vandana, 35
Silent Spring (Carson), 65
Sloman, S., 164
Stanford University, 92
Stelle, Will, 7
Stem cell research, 137–50; analyses, 143–48; history and politics of ESCS, 138–41; overview, 137–38; stem cells, ideology, and values, 141–43
Stem Cell Research Enhancement Act of 2005, 139. *See also* H.R. 810
Stem cells: ideology and, 141–43; values and, 141–43
Synthesis Report Summary for Policymakers, 85

Tank, Jennifer, 33
Technocrats and Democracy, 155
Teen sex, 109–32; abstinence-only education, 116–19; analyses, 123–30; emergency contraception, 114–16; human papillomavirus (HPV), 111–14; ideology, positivism, and postmaterialism, 119–23; overview, 109–11; Plan B, 114–16
Teva Pharmaceutical Industries Ltd., 115
Thimerosal, 57
Three Mile Island nuclear power plant, 2
Title V abstinence-only-until-marriage program, 117
Transgenic crops, 30
Trump, Donald, 62–63, 93

UN Convention of Biological Diversity, 88
United Nations, 34
United Nations Conference on Environment and Development Principle 15, 37
United Nations Environmental Programme, 85
United Nations Framework Convention on Climate Change, 93
University of Wisconsin, 137, 139
Using Science as Evidence in Public Policy report, 153
U.S. Postal Service: Computerized Delivery Sequence file, 13
U.S. public, climate change and, 90–92
U.S. Religious Landscape Study, 109

Vaccines for Children Program, 55
Vaccine skeptics, 60–62
Values, 156–62; genetically modified organisms (GMOs), 35–38; stem cells and, 141–43
Vitamin A deficiency, 34

Wakefield, Andrew, 56
Waltz, E., 33
Washington Post, 155
Whitmarsh, L., 97
WHO, 34, 122
Woodly, Deva R., 166
World Meteorological Organization, 85
World Values Survey, 17
World War II, 11, 97

Yamanaka, Shinya, 149

Zika epidemic, 89

About the Authors

Erika Allen Wolters (PhD, Oregon State University) is the Director of Oregon State University's Policy Analysis Laboratory (OPAL) and a faculty member in Political Science. She holds a PhD in environmental science with much of her research being on environment, science, and policy issues. Current research focuses on the political and ecological linkages between people and the environment, and includes work on water in the Western United States, Consumerism and Sustainability, and Science and Policy. Recent publications include articles in *Social Science Journal, Ocean & Coastal Management*, and several book chapters including a chapter in *Cities, Sagebrush, and Solitude: Urbanization and Cultural Conflict in the Great Basin* (D. Judd and S. Witt, eds., University of Nevada).

Brent S. Steel is Professor and Director of the Public Policy Graduate Program in the School of Public Policy at Oregon State University. He teaches courses in comparative public policy, politics, and administration. Steel is coeditor of *New Strategies for Wicked Problems: Science and Solutions in the 21st Century* (Oregon State University Press) and editor of *Science and Politics: An A-to-Z Guide to Issues and Controversies* (Sage).